THE MESOZOIC

165 MILLION Y...

System / Period	Stage		Age	
	EUROPE		**NORTH AMERICA**	
CRETACEOUS	MAASTRICHTIAN		GULFIAN	
	CAMPANIAN			
	SANTONIAN			
	CONIACIAN			
	TURONIAN			
	CENOMANIAN			
	ALBIAN		COMANCHEAN	
	APTIAN	GARGASIAN / BEDOULIAN		
	BARREMIAN			
	HAUTERIVIAN			
	VALANGINIAN			
	BERRIASIAN			
JURASSIC	TITHONIAN			
	KIMMERIDGIAN			
	OXFORDIAN			
	CALLOVIAN			
	BATHONIAN			
	BAJOCIAN			
	AALENIAN			
	LIASSIC			
TRIASSIC	RHAETIAN			
	NORIAN			
	CARNIAN			
	LADINIAN			
	ANISIAN			
	SCYTHIAN			

MILLIONS YEARS BEFORE THE PRESENT

GREENWOOD LIBRARY
Celebrate Curiosity

This book is in recognition of

Dr. Joseph Garcia

For award of
Emeritus Status
2021

In Suspect Terrain

———

John McPhee

In Suspect Terrain

Farrar · Straus · Giroux

NEW YORK

Copyright © 1982, 1983 by John McPhee
All rights reserved
Printed in the United States of America
Published simultaneously in Canada by
McGraw-Hill Ryerson Ltd., Toronto
Designed by Cynthia Krupat
First printing, 1983

The text of this book originally appeared in
The New Yorker, and was developed with the editorial
counsel of William Shawn, Robert Bingham,
and Sara Lippincott

Geological time scale adapted by Tom Funk from the
third edition of F. W. B. van Eysinga's "Geological
Time Table," Elsevier Scientific Publishing Company,
P.O. Box 211, Amsterdam, The Netherlands

Library of Congress Cataloging in Publication Data
McPhee, John A.   /   In suspect terrain.
1. Geology—Northeastern States.   I. Title.
QE78.3.M36   1983   550   82-21031

To Bill Howarth

In Suspect Terrain

brutal than slow—a continent-to-continent collision marked by an alpine welt, which has reached its old age as the Appalachian Mountains. In the Mesozoic era, two hundred million years ago, rifting began again, pulling apart certain segments of the mountain chain, creating fault-block basins—remnants of which are the Connecticut River Valley, central New Jersey, the Gettysburg battlefields, the Culpeper Basin—and eventually parting the earth's crust enough to start a new ocean, which is now three thousand miles wide and is still growing. Meanwhile, a rhythm of glaciation has been established in what is essentially the geologic present. Ice sheets have been forming on either side of Hudson Bay and have spread in every direction to cover virtually all of Canada, New England, New York, and much of New Jersey, Pennsylvania, and the Middle West. The ice has come and gone at least a dozen times, in cycles that seem to require about a hundred thousand years, and, judging by other periods of glaciation in the earlier history of the earth, the contemporary cycles have only begun. About fifty more advances can be expected. Some geologists have attempted to isolate the time in all time that runs ten thousand years from the Cro-Magnons beside the melting ice to the maternity wards of the here and now by calling it the Holocene epoch, with the implication that this is our time and place, and the Pleistocene—the "Ice Age"—is all behind us. The Holocene appears to be nothing more than a relatively deglaciated interval. It will last until a glacier two miles thick plucks up Toronto and deposits it in Tennessee. If that seems unlikely, it is only because the most southerly reach of the Pleistocene ice fields to date stopped seventy-five miles shy of Tennessee.

Anita Harris is a geologist who does not accept all that is written in that paragraph. She is cool toward aspects of plate tectonics, the novel theory of the earth that explains mountain belts and volcanic islands, ocean ridges and abyssal plains, the deep earthquakes of Alaska and the shallow earthquakes of a fault like the San Andreas as components of a unified narrative, wherein the shell of the earth is divided into segments of varying size, which separate to form oceans, collide to make mountains, and slide by one another causing buildings to fall. In a revolutionary manner, plate-tectonic theory burst forth in the nineteen-sixties, and Anita Harris is worried now that the theory is taught perhaps too glibly in schools. In her words: "It's important for people to know that not everybody believes in it. In many colleges, it's all they teach. The plate-tectonics boys move continents around like crazy. They publish papers every year revising their conclusions. They say that a continental landmass up against the eastern edge of North America produced the Appalachians. I know about some of the geology there, and what they say about it is wrong. I don't say they're wrong everywhere. I'm open-minded. Too often, though, plate tectonics is oversimplified and overapplied. I get all heated up when some sweet young thing with three geology courses tells me about global tectonics, never having gone on a field trip to look at a rock."

As she made these comments, she was travelling west on Interstate 80, approaching Indiana on a gray April morning. She had brought me along to "do geology," as geologists like to say—to see the countryside as she discerned it. Across New Jersey, Pennsylvania, and Ohio, she had been collecting, among other things, limestones and dolomites for their contained conodonts, index fossils from the Paleozoic, whose extraordinary utility in oil and gas exploration had been her discovery, with the result that Mobil and Chevron, Amoco and Arco, Chinese and Norwegians had appeared at her door. She was driving, and she wore a railroad engineer's striped hat, a wool shirt, bluejeans, and old split hiking boots—hydrochloric acid for testing limestones and dolomites in a phial in a case on her hip. With her high cheekbones, her assertive brown eyes, her long dark hair in twin ponytails, she somehow suggested an American aborigine. Of middle height, early middle age, she had been married twice—first to a northern-Appalachian geologist, and now to a southern-Appalachian geologist. She was born on Coney Island and grew up in a tenement in Williamsburg Brooklyn. There was not a little Flatbush in her manner, soul, and speech. Her father was Russian, and his name in the old country was Herschel

Litvak. In Brooklyn, he called himself Harry Fishman, and sometimes Harry Block. According to his
daughter, English names meant nothing to Russian
Jews in Brooklyn. She grew up Fishman and became
in marriage Epstein and Harris, signing her geology
with her various names and imparting some difficulty
to followers of her professional papers. With her
permission, I will call her Anita, and let the rest of
the baggage go. Straightforwardly, as a student, she
went into geology because geology was a means of
escaping the ghetto. "I knew that if I went into geology I would never have to live in New York City,"
she once said to me. "It was a way to get out." She
was nineteen years old when she was graduated from
Brooklyn College. She remembers how pleased and
astounded she was to learn that she could be paid
"for walking around in mountains." Paid now by the
United States Geological Survey, she has walked un-
counted mountains.

After the level farmlands of northwestern Ohio,
the interstate climbed into surprising terrain—surprising enough to cause Anita to suspend her attack
on plate tectonics. Hills appeared. They were steep
in pitch. The country resembled New England, a
confused and thus beautiful topography of forested
ridges and natural lakes, stone fences, bunkers and
bogs, cobbles and boulders under maples and oaks:
Indiana. Rough and semi-mountainous, this corner of
Indiana was giving the hummocky lie to the reputed
flatness of the Middle West. Set firmly on the craton

—the Stable Interior Craton, unstirring core of the continent—the whole of Middle America is structurally becalmed. Its basement is coated with layers of rock that are virtually flat and have never experienced folding, let alone upheaval. All the more exotic, then, were these abrupt disordered hills. Evidently superimposed, they almost seemed to have been created by the state legislature to relieve Indiana. Not until the nineteenth century did people figure out whence such terrain had come, and how and why. "Look close at those boulders and you'll see a lot of strangers," Anita remarked. "Red jasper conglomerates. Granite gneiss. Basalt. None of those are from anywhere near here. They're Canadian. They have been transported hundreds of miles."

The ice sheets of the present era, in their successive spreadings overland, have borne immense freight—rock they pluck up, shear off, rip from the country as they move. They grind much of it into gravel, sand, silt, and clay. When the ice melts, it gives up its cargo, dumping it by the trillions of tons. The most recent advance has been called the Wisconsinan ice sheet, because its effects are well displayed in Wisconsin. Its effects, for all that, are not unimpressive in New York. The glacier dumped Long Island where it is (nearly a hundred per cent of Long Island), and Nantucket, and Cape Cod, and all but the west end of Martha's Vineyard. Wherever the ice stopped and began to melt back, it signed its retreat with terminal moraines—huge accumulations of undifferentiated rock, sand, gravel, and clay. The ice stopped at Perth Amboy, Metuchen, North Plainfield, Madison, Morristown—leaving a sinuous, morainal, lobate line that not only connects these New Jersey towns but keeps on going to the Rocky Mountains. West of Morristown, old crystalline rock from the earth's basement—long ago compressed, distorted, and partially melted, driven upward and westward in the Appalachian upheavals—stands now in successive ridges, which are called the New Jersey Highlands. They trend northeast-southwest. With a notable exception, they have discouraged east-west construction of roads. When the last ice sheet set down its terminal moraine, it built causeways from one ridge to another, on which Interstate 80 rides west. Over the continent, the ice had spread southward about as evenly as spilled milk, and there is great irregularity in its line of maximum advance. South of Buffalo, it failed to reach Pennsylvania, but it plunged deep into Ohio, Indiana, Illinois. The ice sheets set up and started Niagara Falls. They moved the Ohio River. They dug the Great Lakes. The ice melted back in stages. Pausing here and there in temporary equilibrium, it sometimes readvanced before continuing its retreat to the north. Wherever these pauses occurred, as in northeastern Indiana, boulders and cobbles and sand and gravel piled up in prodigal quantity—a cadence of recessional moraines, hills of rock debris. The material, heterogeneous and unsorted, has its own style of fabric, in which many hundreds of Suiattles and Yentnas, most of which are gone now, leaving their works behind. The rivers have built outwash plains beyond the glacial fronts, sorting and smoothing miscellaneous sizes of rock—moving cobbles farther than boulders, and gravels farther than cobbles, and sands farther than gravels, and silt grains farther than sands—then gradually losing power, and filling up interstices with groutings of clay. Enormous chunks of ice frequently broke off the retreating glaciers and were left behind. The rivers built around them containments of gravel and clay. Like big, buried Easter eggs, the ice sat there and slowly melted. When it was gone, depressions were left in the ground, pitting the outwash plains. The depressions have the shapes of kettles, or at least have been so described, and "kettle" is a term in geology. All kettles contained water for a time, and some contain water still. Rivers that developed under glaciers ran in sinuous grooves. Rocks and boulders coming out of the ice fell into the rivers, building thick beds contained between walls of ice. When the glacier was gone, the riverbeds were left as winding hills. The early Irish called them eskers, meaning pathways, because they used them as means of travel above detentive bogs. Where debris had been concentrated in glacial crevasses, melting ice left hillocks, monticles, hummocks, knolls, braes—collections of lumpy hills known generically to the Scots as kames. In Indiana as in Scotland—in La Bresse and Estonia as in New

England and Quebec—the sort of country left behind after all these features have been created is known as kame-and-kettle topography.

The interstate was waltzing with the glacier—now on the outwash plain, now on moraine, among the kettles and kames of Scottish Indiana. Roadcuts were green with vetch covering glacial till. We left 80 for a time, the closer to inspect the rough country. The glacier had been away from Indiana some twelve thousand years. There were many beds of dried-up lakes, filled with forest. In the Boundary Waters Area of northern Minnesota, the ice went back ten thousand years ago, possibly less, and most of the lakes it left behind are still there. The Boundary Waters Area is the scene of a contemporary conservation battle over the use and fate of the lakes. "Another five thousand years and there won't be much to fight about," Anita said, with a shrug and a smile. "Most of those Minnesota lakes will probably be as dry as these in Indiana." Some of the larger and deeper ones endure. We made our way around the shores of Lake James, Bingham Lake, Lake of the Woods, Loon Lake. Like Walden Pond, in Massachusetts, they were kettles.

The woods around them were bestrewn with boulders, each an alien, a few quite large. If a boulder rests above bedrock of another type, it has obviously been carried some distance and is known as an erratic. In Alaska, I have come upon glacial

erratics as big as office buildings, with soil developing on their tops and trees growing out of them like
hair. In Pokagon State Park, Indiana, handsome
buildings looked out on Lake James—fieldstone
structures, red and gray, made of Canadian rocks.
The red jasper conglomerates were from the north
shore of Lake Huron. The banded gray gneisses were
from central Ontario. The sources of smaller items
brought to Indiana by the ice sheets have been less
easy to trace—for example, diamonds and gold. During the Great Depression, one way to survive in Indiana was to become a pick-and-shovel miner and
earn as much as five dollars a day panning gold from
glacial drift—as all glacial deposits, sorted and unsorted, are collectively called. There were no nug-
gets, nothing much heavier than a quarter of an
ounce. But the drift could be fairly rich in fine gold.
It had been scattered forth from virtually untrace-
able sources in eastern Canada. One of the oddities
of the modern episodes of glaciation is that while
three-fifths of all the ice in the world covered North
America and extended south of Springfield, Illinois,
the valley of the Yukon River in and near Alaska
was never glaciated, and as a result the gold in the
Yukon drainage—the gold of the richest placer
streams ever discovered in the world—was left
where it lay, and was not plucked up and similarly
scattered by overriding ice. Miners in Indiana
learned to look in their pans for menaccanite—bean-
like pebbles of iron and titanium that signalled with

the materials are diamonds. Evidently, there are no diamond pipes, as they are also called, in or near Indiana. Like the huge red jasper boulders and the tiny flecks of gold, Indiana's diamonds are glacial erratics. They were transported from Canada, and by reading the fabric of the till and taking bearings from striations and grooves in the underlying rock —and by noting the compass orientation of drumlin hills, which look like sculptured whales and face in the direction from which their maker came—anybody can plainly see that the direction from which the ice arrived in this region was something extremely close to 045°, northeast. At least one pipe containing gem diamonds must exist somewhere near a line between Indianapolis and the Otish Mountains of Quebec, because the ice that covered Indiana did not come from Kimberley—it formed and grew and, like an opening flower, spread out from the Otish Mountains. With rock it carried and on rock it traversed, it narrated its own journey, but it did not reveal where it got the diamonds.

There is a layer in the mantle, averaging about sixty miles below the earth's surface, through which seismic tremors pass slowly. The softer the rock, the slower the tremor—so it is inferred that the low-velocity zone, as it is called, is partly fluid. In the otherwise solid mantle, it is a level of lubricity upon which the plates of the earth can slide, interacting at their borders to produce the effects known as plate tectonics. The so-termed lithospheric plates, in other words, consist of crust and uppermost mantle and can be as much as ninety miles thick. Diamond pipes are believed to originate a good deal deeper than that—and in a manner which, as most geologists would put it, "is not well understood." After drawing fuel from surrounding mantle rock—compressed water from mica, in all likelihood, and carbon dioxide from other minerals—the material is thought to work slowly upward into the overlying plate. Slow it may be at the start, but a hundred and twenty miles later it comes out of the ground at Mach 2. The result is a modest crater, like a bullet hole between the eyes.

No one has ever drilled a hundred and twenty miles into the earth, or is likely to. Diamond pipes, meanwhile, have brought up samples of what is there. It is spewed all over the landscape, but it also remains stuck in the throat, like rich dense fruitcake. For the most part, it is peridotite, which is the lowest layer of the subcontinental package and is suspected to be the essence of the mantle. There is high-pressure recrystallized basalt, full of garnets and jade. There are olivine crystals of incomparable size. The whole of it is known as kimberlite, the matrix rock of diamonds.

The odds against diamonds appearing in any given pipe are about a hundred to one. Carbon will crystallize in its densest form only under conditions of considerable heat and pressure—pressures of the sort that exist deep below the thickest parts of the plates, pressures of at least a hundred thousand pounds per square inch. The thickest parts of the plates are the continental cores, the cratons. All diamond-bearing kimberlites ever found have been in pipes that came up through cratons. Down where diamonds form, they are stable, but as they travel upward they pass through regions of lower pressure, where they will swiftly turn into graphite. Only by passing through such regions at tremendous speed can diamonds reach the earth's surface as diamonds, where they cool suddenly and enter a state of precarious preservation that somehow betokens to human beings a touching sense of "forever." Diamonds shoot like bullets through the earth's crust. Nonetheless, they are often found within rinds of graphite. Countless quantities turn into graphite altogether or disappear into the air as carbon dioxide. At room temperature and surface pressure, diamonds are in repose on an extremely narrow thermodynamic shelf. They want to be graphite, and with a relatively modest boost of heat graphite is what they would become, if atmospheric oxygen did not incinerate them first. They are, in this sense, unstable—these finger-flashing symbols of the eternity of vows, yearning to become fresh pencil lead. Except for particles that are sometimes found in meteorites, diamonds present themselves in nature in no other way.

Kimberlite is easily eroded. A boy playing jacks in South Africa in 1867 picked up an alluvial diamond that led to the discovery of a number of pipes,

Canada to look for a diamond pipe, but when you have diamonds in this drift you'd better believe it is telling you that diamond pipes are there. Rocks are the record of events that took place at the time they formed. They are books. They have a different vocabulary, a different alphabet, but you learn how to read them. Igneous rocks tell you the temperature at which they changed from the molten to the solid state, and they tell you the date when that happened, and hence they give you a picture of the earth at that time, whether they formed three thousand million years ago or flowed out of the ground yesterday. In sedimentary rock, the colors, the grain sizes, the ripples, the crossbedding give you clues to the energy of the environment of deposition—for example, the force and direction and nature of the rivers that laid down the sediments. Tracks and trails left by organisms—and hard parts of their bodies, and flora in the rock—tell whether the material came together in the ocean or on the continent, and possibly the depth and temperature of the water, and the temperature on the land. Metamorphic rocks have been heated, compressed, and recrystallized. Their mineral composition tells you if they were originally igneous or sedimentary. Then they tell you what happened later on. They tell you the temperatures when they changed. At one point, I wanted to major in history. My teachers steered me into science, but I really majored in history. I grew up in topography like this, believe it or not. Looking at or no old neighborhood, he said, he would not go near Williamsburg, or for that matter a good many other places in Brooklyn; and he reeled off stories of open carnage that might have tested the stomach of the television news. I wondered what it might be like to die defending myself with a geologist's rock hammer. Anita, for her part, seemed nervous as we left for the city. Twenty-five years away, she seemed afraid to go home.

It was an August day already hot at sunrise. "In Williamsburg, I lived at 381 Berry Street," she said as we crossed the big bridge. "It was the worst slum in the world, but the building did have indoor plumbing. Our first apartment there was a sixth-floor walkup. The building was from the turn of the century and was faced with red Triassic sandstone." Brooklyn was spread out before us, and Manhattan stood off to the north, with its two sets of skyscrapers three miles apart—the ecclesiastical spires of Wall Street, and beyond them the midtown massif. Anita asked me if I had ever wondered why there was a low saddle in the city between the stands of tall buildings.

I said I had always assumed that the skyline was shaped by human considerations—commercial, historical, ethnic. Who could imagine a Little Italy in a skyscraper, a linoleum warehouse up in the clouds?

The towers of midtown, as one might imagine, were emplaced in substantial rock, Anita said—rock that once had been heated near the point of melting,

had recrystallized, had been heated again, had recrystallized, and, while not particularly competent,
was more than adequate to hold up those buildings. Most important, it was right at the surface. You
could see it, in all its micaceous glitter, shining like
silver in the outcrops of Central Park. Four hundred
and fifty million years in age, it was called Manhattan schist. All through midtown, it was at or near the
surface, but in the region south of Thirtieth Street
it began to fall away, and at Washington Square it
descended abruptly. The whole saddle between midtown and Wall Street would be underwater, were it
not filled with many tens of fathoms of glacial till. So
there sat Greenwich Village, SoHo, Chinatown, on
material that could not hold up a great deal more
than a golf tee—on the ground-up wreckage of the
Ramapos, on crushed Catskill, on odd bits of Nyack
and Tenafly. In the Wall Street area, the bedrock
does not return to the surface, but it comes within
forty feet and is accessible for the footings of the
tallest things in town. New York grew high on the
advantage of its hard rock, and, New York being
what it is, cities all over the world have attempted to
resemble New York, in much the way that golf
courses all around the world have attempted to resemble St. Andrews. The skyline of nuclear Houston,
for example, is a simulacrum of Manhattan's. Houston rests on twelve thousand feet of montmorillonitic
clay, a substance that, when moist, turns into mobile
jelly. After taking so much money out of the ground,

the oil companies of Houston have put hundreds of millions back in. Houston is the world's foremost city in fat basements. Its tall buildings are magnified duckpins, bobbing in their own mire.

We skirted Brooklyn on the Belt Parkway, heading first for Coney Island, where Anita had spent many a day as a child, and where, somewhat impatiently, she had been born. Her mother, seven months pregnant, took a subway to the beach one day, and Anita first drew breath in Coney Island Hospital.

"Cropsey Avenue," she said now, reading a sign. "Keep right, we're going off here."

I went into the right lane, signals blinking, but the exit was chocked with halted traffic. There were police. There were flashing lights. Against the side of an abused Pontiac, an evidently unruly young man was leaning palms flat, like a runner stretching, while a cop addressed him with a drawn pistol. "Welcome home, Anita," said Anita.

The broad beach was silent, so early in the morning, where people in ten thousands had been the day before, and where numbers just as great would soon return. The Parachute Jump stood high in relief. The Cyclone was in shadow and touched by slanting light. Reminiscently, Anita ran her eye from the one to the other and to the elevated railways beyond. When a fossil impression is left in sand by the outside of an organic structure, it is known in geology as an external mold. One would not have to be a sedimentologist to read this beach, with its colonies of giant bivalves. We walked to the strand-line, the edge of the water, where the play of waves had concentrated heavy dark sands—hematite, mag-netite, small garnets broken out by the glacier from their matrix of Manhattan schist.

The beach itself, with its erratic sands, was the extremity of the outwash plain. The Wisconsinan ice sheet, arriving from the north, had come over the city not from New England, as one might guess, but primarily from New Jersey, whose Hudson River counties lie due north of Manhattan. Big boulders from the New Jersey Palisades are strewn about in Central Park, and more of the same diabase is scat-tered through Brooklyn. The ice wholly covered the Bronx and Manhattan, and its broad snout moved across Astoria, Maspeth, Williamsburg, and Bedford-Stuyvesant before sliding to a stop in Flatbush. Flat-bush was the end of the line, the point of return for the "Ice Age," the locus of the terminal moraine. Water poured in white tumult from the melting ice, carrying and sorting its freight of sands and gravels, building the outwash plain: Bensonhurst, Canarsie, the Flatlands, Coney Island. When Anita was a child, she would ride the D train out to Coney Is-land, with an old window screen leaning against her knees. She sifted the beach sand for lost jewelry. In the beach sand now, she saw tens of thousands of garnets. There is a lot of iron in the Coney Island beach as well, which makes it tawny from oxidation,

and not a lot of quartz, which would make it white.
The straw-colored sand sparkled with black and silver micas—biotite, muscovite—from Fifth Avenue
or thereabouts, broken out of Manhattan schist. A
beach represents the rock it came from. Most of
Coney Island is New Jersey diabase, Fordham
gneiss, Inwood marble, Manhattan schist. Anita
picked up some sand and looked at it through a hand
lens. The individual grains are characteristically
angular and sharp, she said, because the source rock
was so recently crushed by the glacier. To make a
well-rounded grain, you need a lot more time.
Weather and waves had been working on this sand
for fifteen thousand years.

If the gneissic grains and garnets were erratics,
so in their way were the Schenley bottles, the Pepsi-
Cola cans, the Manhattan Schlitz, the sand-coated
pickles and used paper plates.

"Colonial as penguins, dirtier than mud dau-
bers," I observed of the creatures of the beach.

"We rank with bats, starlings, and Pleistocene
sloths as the great messmakers of the world," said
Anita, and we left Coney Island for Williamsburg.

North over the outwash plain we followed Ocean
Parkway five miles—broad, tree-lined Ocean Parkway, with neat houses in trim neighborhoods, reaching into shaded streets. Ahead, all the while, loomed
the terminal moraine, suggesting, from a distance, an
escarpment, but actually just a fairly steep hill. Eastern Parkway defines its summit, two hundred feet

high. Two hundred feet of till. Near Prospect Park you begin to climb. One moment you are level on the plain and the next you are nose up, gaining altitude. There are cemeteries in every direction: Evergreens Cemetery, Lutheran Cemetery, Mt. Carmel, Cypress Hills, Greenwood Cemetery—some of the great necropolises of all time, with three million under sod, moved into the ultimate neighborhood, the terminal moraine. "In glacial country, all you have to do is look for cemeteries if you want to find the moraine," Anita said. "A moraine is poor farmland—steep and hummocky, with erratics and boulders. Yet it's easy ground to dig in, and well drained. An outwash plain is boggy. There's a cemetery over near Utica Avenue that's in the outwash. Most people prefer moraine. I would say it's kind of distasteful to put your mother down into a swamp."

Ebbets Field, where they buried the old Brooklyn Dodgers, was also on the terminal moraine. When a long-ball hitter hit a long ball, it would land on Bedford Avenue and bounce down the morainal front to roll toward Coney on the outwash plain. No one in Los Angeles would ever hit a homer like that.

We detoured through Prospect Park, which is nestled into the morainal front and is studded with big erratics on raucously irregular ground. It looks much like Pokagon Park, in Indiana, with the difference that the erratics there are from the Canadian Shield and these were from the New Jersey Palisades. Pieces of the Adirondacks have been found in Pennsylvania, pieces of Sweden on the north German plains, and no doubt there is Ticonderoga dolomite, Schenectady sandstone, and Peekskill granite in the gravels of Canarsie and the sands of Coney Island. But such distant transport, while it characterizes continental ice sheets wherever they have moved, accounts for a low percentage of the rock in glacial drift. The glacier cuts and fills. Continuously, it plucks up material and sets it down, plucks it up, sets it down. It taketh away, and then it giveth. A diamond may travel from Quebec to Indiana, some dolomite from Lake George to the sea, but most of what is lifted is dropped nearby—boulders from New Jersey in Prospect Park.

"Glacial geology is simple to deal with," Anita said, "because so much of what the glacier created is preserved. Also, you can go places and see the same processes working. You can go to Antarctica and see continental glaciation. There's alpine glaciation in Alaska."

This warm clear summer day was now approaching noon, and Prospect Park was quiet and unpeopled. It was all but deserted. Anita as a child had come here often. She remembered people and picnics everywhere she looked, none of this ominous silence. "I suppose it isn't safe," she said, and we moved on toward Williamsburg.

As we drew close, she became even more obviously nervous. "They tell me it's just the worst slum in the world now," she said. "I don't know if I should tell you to roll up all the windows and lock the doors."

"We would die of the heat."

"This is a completely unnatural place," she went on. "It's a totally artificial environment. Cockroaches, rats, human beings, and pigeons are all that survive. At Brooklyn College, my instructors had difficulty relating geology to the lives of people in this artificial world. In the winter, maybe you froze your ass off waiting for the subway. Maybe that was a way to begin discussing glaciation. In the city, let me tell you, no one knows from geology."

We went first to her high school. It appeared to be abandoned and was not. It was a besooted fortress with battlements. Inside were tall cool hallways that smelled of polish and belied the forbidding exterior. She had walked the halls four years with A's on her report cards and been graduated with high distinction at the age of fifteen. We went to P.S. 37, her grade school. It was taller than wide and looked like an old brick church. It was abandoned, beyond a doubt—glassless and crumbling. Trees of heaven, rooted in the classroom floors, were growing out the windows. Anita said, "At least I'm glad I saw my school, I think, before they take it away."

We came to Broadway and Berry Street, and now she had before her for the first time in twenty-five years the old building where she had lived. It was a six-story cubical tenement, with so many fire escapes that it seemed to be faced more with iron than with the red Triassic stone. Anita looked at the building in silence. Usually quick to fill the air with words, she said nothing for long moments. Then she said, "It doesn't look as bad as it did when I lived here."

She stared on at the building for a while before speaking again, and when she did speak the nervousness of the morning was completely gone from her voice. "It's been sandblasted," she said. "They've cleaned it up. They've put a new facing on the lower stories, and they've sandblasted the whole building. People are wrong. They're wrong in what they tell me. This place looks cleaner than when I lived here. The whole neighborhood still looks all right. It hasn't changed. I used to play stickball here in the street. This is my neighborhood. This is the same old neighborhood I grew up in. I'm not afraid of this. I'm getting my confidence up. I'm not afraid."

We moved along slowly from one block to another. A young woman crossed the street in front of us, pushing a baby carriage. "She's wearing a wig, I promise you," Anita said. "Her head may be shaved." Singling out another woman among the heterogeneous people of the neighborhood, she said, "Look. See that woman with the turban? She has her hair covered on purpose. They're Chassidic Jews. Their hair is shaved off or concealed so they will not be attractive to passing men." There was a passing man with long curls hanging down either side of his head—in compliance with a dictum of the Pentateuch. "Just to be in the streets here is like stepping into the Middle Ages," Anita said. "Fortunately, my parents were not religious. I would have thought these people would have moved out of here long ago. Chassidic Jews are not all poor, I promise you. Their houses may not look like much, but you should see them inside. They're diamond-cutters. They handle money. And they're still here. People are wrong. They are wrong in what they have told me."

We went out of the noon sun into deep shade under the Williamsburg Bridge, whose immense stone piers and vaulting arches seemed Egyptian. She had played handball under there when she was a girl. "There were no tennis courts in this part of the world, let me tell you." When the boys went off to swim in the river, she went back to Berry Street. "Me? In the river? Not me. The boys swam nude."

In the worst parts of summer, when the air was heavy and the streets were soft, Anita went up onto the bridge, climbing to a high point over the river, where there was always a breeze. Seven, eight years old, she sat on the pedestrian walk, with her feet dangling, and looked down into the Brooklyn Navy Yard. The Second World War was in full momentum. U.S.S. Missouri, U.S.S. Bennington, U.S.S. Kearsarge —she saw keels going down and watched battleships and carriers grow. It was a remarkable form of entertainment, but static. Increasingly, she wondered what lay beyond the bridge. One day, she got up the courage to walk all the way across. She set foot on Manhattan and immediately retreated. "I wanted to go up Delancey Street, but I was too scared."

Next time, she went up Delancey Street three blocks before she turned around and hurried home. In this manner, through time, she expanded her horizons. In the main, she just looked, but sometimes she had a little money and went into Manhattan stores. About the only money she ever had she earned returning bottles for neighbors, who gave her a percentage of the deposit. Her idea of exceptional affluence was a family that could afford fresh flowers. Her mother was a secretary whose income covered a great deal less than the family's needs. Her father was a trucker ("with a scar on his face that would make you think twice"), and his back had been broken in an accident. He would spend three years in traction, earning nothing. Gradually, Anita's expeditions on foot into Manhattan increased in length until she was covering, round trip, as much as twelve miles. Her line of maximum advance was somewhere in Central Park. "That's as far as I ever got. I was too scared." Going up the Bowery and through the East Village, she had no more sense of the geology than did the men who were lying in the doorways. When she looked up at the Empire State Building, she was unaware that it owed its elevation to the formation that outcropped in Central Park; and when she saw the outcrops there, she did not wonder why, in the moist atmosphere of the American East, those great bare shelves of sparkling rock were not covered with soil and vegetation. In Wyoming, wind might have stripped them bare, but Wyoming is miles high and drier than the oceans of the moon. Here in the East, a river could wash rock clean, but this rock was on the high ground of an island, far above flood and tide. She never thought to wonder why the rock was scratched and grooved, and elsewhere polished like the foyer of a bank. She didn't know from geology.

In Brooklyn College, from age fifteen onward, she read physics, mineralogy, structural geology, igneous and metamorphic petrology. She took extra courses to the extent permitted. To attend the college she had to pay six dollars a semester, and she meant to get everything out of the investment she could. There were also lab fees and breakage fees. Breakage fees, in geology, were not a great problem. Among undergraduate colleges in the United States, this one was relatively small, about the size of Harvard, which it resembled, with its brick-and-white-trim sedate Colonial buildings, its symmetrical courtyards and enclosed lawns; and like Harvard it stood on outwash. Brooklyn College is in south Flatbush, seaward of the terminal moraine. When Anita was there, in the middle nineteen-fifties, there were so many leftists present that the college was known as the Little Red Schoolhouse. She did not know from politics, either. She was in a world of roof pendants and discordant batholiths, elastic collisions and neutron scatteration, and she branched out into mineral deposits, field mapping, geophysics, and historical geology, adding such things to the skills she had established earlier in accounting, bookkeeping, typing, and shorthand. It had been assumed in her family that she would be a secretary, like her mother.

Now when she goes up Fifth Avenue—as she did with me that summer day—she addresses Fifth Avenue as the axis of the trough of a syncline. She knows what is underfoot. She is aware of the structure of the island. The structure of Manhattan is one of those paradoxes in spatial relations which give geologists especial delight and are about as intelligible to everyone else as punch lines delivered in Latin. There is a passage in the œuvre of William F. Buckley, Jr., in which he remarks that no writer in the history of the world has ever successfully made clear to the layman the principles of celestial navigation. Then Buckley announces that celestial navigation is dead simple, and that he will pause in the development of his present narrative to redress forever the failure of the literary class to elucidate this abecedarian technology. There and then—and with intrepid, awesome courage—he begins his explication; and before he is through, the oceans are in orbit, their barren shoals are bright with shipwrecked stars. With that preamble, I wish to announce that I am about to make perfectly clear how

Fifth Avenue, which runs along the high middle of a loaf of rock that lies between two rivers, runs also up the center of the trough of a syncline. When rock is compressed and folded (like linen pushed together on a table), the folds are anticlines and synclines. They are much like the components of the letter S. Roll an S forward on its nose and you have to the left a syncline and to the right an anticline. Each is a part of the other. Such configurations in rock compose the structure of a region, but will not necessarily shape the surface of the land. Erosion is the principal agent that shapes the surface of the land; and erosion—particularly when it packs the violence of a moving glacier—can cut through structure as it pleases. A carrot sliced the long way and set flat side up is composed of a synclinal fold. Manhattan, embarrassingly referred to as the Big Apple, might at least instructively be called the Big Carrot. River to river, erosion has worn down the sides, and given the island its superficial camber. Fifth Avenue, up on the high ground, is running up the center of a synclinal trough.

On the upper West Side that afternoon, Anita drew her rock hammer and relieved Manhattan of some dolomite marble, which she took from an outcrop for its relevance to her research in conodonts. She found the marble "overcooked." She said, "To get that kind of temperature, you have to go down thirty or forty thousand feet, or have molten rock nearby, or have a high thermal gradient, which can vary from place to place on earth by a factor of four. This marble is so cooked it is almost volatilized. This —you better believe—is hot rock." At Seventy-second Street and West End Avenue, she stopped to admire a small apartment building whose façade, in mottled greens and black, was elegant with serpentine. On Sixty-eighth Street between Fifth and Madison, she was impressed by a house of gabbro, as anyone would be who had spent a childhood emplaced like a fossil in Triassic sand. It was a house of great wealth, the house of gabbro. Up the block was a house of granite, even grander than the gabbro, and beyond that was a limestone mansion so airily patrician one feared it might dissolve in rain. Anita dropped acid on it and watched it foam.

. . .

Jack Epstein, Anita's northern-Appalachian geologist, went to Brooklyn College, too, and subsequently enrolled in the master's program at the University of Wyoming. Anita tried to follow, in 1957, but the geology department in Laramie offered no fellowships for first-year graduate students. ("I needed money. I didn't have a pot to cook in.") She looked into places like Princeton, with geology departments outstanding in the world, but they were even less receptive than Wyoming. In those days, Princeton would not have admitted a woman had she been a direct descendant of Sir Charles Lyell offer-

ing as tuition her weight in gems. Anita applied to
ten schools in all. The best offer came from Indiana
University, in Bloomington, where her professors
were soon much aware of her as an extremely bright
and aggressive student with the disconcerting habit
of shaking her head while they talked, as if to say no,
no, no, no, you cratonic schnook, you don't know
from nothing. Something of the sort was not always
far from her thoughts. ("I am not a very orthodox
geologist. I do buy some dogma, if I think it's com-
mon sense.")

Bloomington stood upon Salem limestone,
which, in the terminology of the building trade,
makes beautiful "dimension stone," and is cut to be
the cladding of cities. It formed from lime mud in
the Meramecian age of late Mississippian time—
between 332 and 327 million years before the present
—when Bloomington was at the bottom of a shallow
arm of the transgressing ocean, an epicratonic sea.
"You people in New York may have your Empire
State Building," a professor pointed out to Anita.
"But out here we have the hole in the ground it came
from."

Anita and Jack Epstein were married in 1958,
and, with their newly acquired master's degrees,
went to work for the United States Geological Survey. Within the profession, the Survey had particular
prestige. A geologist who sought field experience was
likely to obtain it in such quantity and variety nowhere else. Anita and Jack Epstein looked upon

geology as "an extremely applied science" and shared a conviction that field experience was indispensable in any geological career—no less essential to a modern professor than it ever was to a pick-and-shovel prospector. ("People should go out and get experience and not just turn around and teach what they've been taught.") In their first year in the Survey—to an extent beyond anything they could ever have guessed—they would get what they sought.

Because geology is sometimes intuitive even to the point of being subjective, the sort of field experience one happens to acquire may tend to influence one's posture with regard to deep questions in the science. Geologists who grow up with young rocks are likely to subscribe strongly to the doctrine of uniformitarianism, whereby the present is seen to be the key to the past. They discern a river sandbar in a wall of young rock; they see a sandbar in a living river; and they know that each is in the process of becoming the other, cyclically through time. Whatever is also was, and ever again shall be. Geologists who grow up with very old rock tend to be impressed by the fact that it has been around since before the earliest development of life, and to imagine a progression in which the recycling of the earth's materials is a subplot in a dramatic story that begins with dark scums in motion on an otherwise featureless globe and evolves through various continental configurations toward the scenery of the earth today.

They refer to the earliest part of that story as "scum tectonics." The rock cycle—with its crumbling mountains being carried to the sea to form there the rock of mountains to be—is the essence of the uniformitarian principle, which was first articulated by James Hutton, of Edinburgh, at the end of the eighteenth century, marking the beginning of modern geological insight and the decline of the theological notion that the earth is a few thousand years old and that man has been a participant in its history almost from the beginning. Radiometrics and any number of other cross-checking measurements of time now tell us that the earth existed about forty-six hundred million years before the beginning of the Judeo-Christian era, thereby recasting mankind as an *arriviste* species, an obviously unsettled obtruder.

Before Hutton, geology was seen as a succession of catastrophes—most notably Noah's Flood, which not only had placed oysters, clams, and other marine fossils in mountain rock but had sculpted most of the features of the modern earth. All else was "antediluvian." The world had been shaped by brief, cataclysmic events. Hutton, with his depths of time —his vision of great crustal changes occurring slowly through unguessable numbers of years—opened the way to Darwin (time is the first requirement of evolution) and also placed emphasis on repetitive processes and a sense that change is largely gradual. In contemporary dress, these concepts are still at odds in geology. Some geologists seem to look upon the rock record as a frieze of catastrophes interspersed with gaps, while others prefer to regard everything from rockslides and volcanic eruptions to rifted continents and plate collisions as dramatic passages in a quietly unfolding story. If you grow up in Brooklyn, you are free to form your prejudices where you may.

Anita Epstein's sense of the dynamics of the earth underwent considerable adjustment one night in 1959, when she and her husband were on summer field assignment in southwestern Montana. They were there to do geologic mapping and studies in structure and stratigraphy in the Madison Range and the Gallatin Range, where Montana is wrapped around a corner of Yellowstone Park. They lived in a U.S.G.S. house trailer in a grove of aspens on the Blarneystone Ranch, a lovely piece of terrain whose absent owner was Emmett J. Culligan, the softener of water. Since joining the Survey, they had worked in Pennsylvania, mapping quadrangles in the region of the Delaware Water Gap, and had spent the winter at headquarters in Washington, and now they were being given a chance to see some geology in a part of the United States where it is particularly visible—in Anita's words, "where it all hangs out."

The ranch was close by Hebgen Lake, which owed itself to a dam in the valley of the Madison River. The valley ran along the line of a fault that was thought to be inactive until that night. The air was crisp. The moon was full. The day before, a fire watcher in a tower in the Gallatins had become aware of an unnerving silence. The birds were gone, he realized. Birds of every sort had made a wholesale departure from his mountain. It would be noted by others that bears had taken off as well, while bears that remained walked preoccupied in circles. The Epsteins had no knowledge of these signs and would not have known what to make of them if they had. They were unaware then that Chinese geologists routinely watch wildlife for intimations of earthquakes. They were also unaware that David Love, of the Survey's office in Laramie, had published an abstract only weeks before called "Quaternary Faulting in and near Yellowstone Park," in which he expressed disagreement with the conventional wisdom that seismic activity on a grand scale was a thing of the past in that region. He said he thought a major shock was not unlikely. Anita was shuffling cards, 11:37 P.M., when the lantern above her began to swing, crockery fell from cabinets, and water leaped out of a basin. Jack tried to catch the swinging lantern and "it beaned him on the head." The floor of the trailer was moving in a way that reminded her of the Fun House at Coney Island. They ran outside. "Trees were toppling over. The solid earth was like a glop of jelly," she would recall later. In the moonlight, she saw soil moving like ocean waves, and for all her professed terror she was collected enough to notice that the waves were not propagating well and were cracking at their crests. She remembers something like thirty seconds of "tremendous explosive noise,"

an "amplified tornado." She was close to the epi-
center of a shock that was felt three hundred and
fifty miles away and markedly affected water wells in
Hawaii and Alaska. East and west from where she
stood ran an eighteen-mile rip in the surface of the
earth. The fault ran straight through Culligan's
ranch house, and had split its levels, raising the back
twelve feet. The tornado sound had been made by
eighty million tons of Precambrian mountainside,
whose planes of schistosity had happened to be inclined toward the Madison River, with the result
that half the mountain came falling down in one of
the largest rapid landslides produced by an earthquake in North America in historical time. People
were camped under it and near it. Among the dead
were some who died of the air blast, after flapping
like flags as they clung to trees. Automobiles rolled
overland like tumbleweed. They were inundated as
the river pooled up against the rockslide, and they
are still at the bottom of Earthquake Lake, as it is
called—a hundred and eighty feet deep.

The fault offset the water table, and the consequent release of artesian pressure sent grotesque
fountains of water, sand, and gravel spurting into the
air. Yet the dam at Hebgen Lake held—possibly because the lake's entire basin subsided, in places as
much as twenty-two feet. Seiche waves crossed its
receding surface. A seiche is a freshwater tsunami,
an oscillation in a bathtub. The surface of Hebgen
Lake was aslosh with them for twelve hours, but the

first three or four were the large ones. Entering lake-side bungalows, they drowned people in their beds.

When a volcano lets fly or an earthquake brings down a mountainside, people look upon the event with surprise and report it to each other as news. People, in their whole history, have seen compara-tively few such events; and only in the past couple of hundred years have they begun to sense the patterns the events represent. Human time, regarded in the perspective of geologic time, is much too thin to be discerned—the mark invisible at the end of a ruler. If geologic time could somehow be seen in the perspec-tive of human time, on the other hand, sea level would be rising and falling hundreds of feet, ice would come pouring over continents and as quickly go away. Yucatáns and Floridas would be under the sun one moment and underwater the next, oceans would swing open like doors, mountains would grow like clouds and come down like melting sherbet, con-tinents would crawl like amoebae, rivers would ar-rive and disappear like rainstreaks down an um-brella, lakes would go away like puddles after rain, and volcanoes would light the earth as if it were a garden full of fireflies. At the end of the program, man shows up—his ticket in his hand. Almost at once, he conceives of private property, dimension stone, and life insurance. When a Mt. St. Helens as-saults his sensibilities with an ash cloud eleven miles high, he writes a letter to the New York *Times* rec-ommending that the mountain be bombed.

As the night returned to quiet and the ground

ceased to move, Anita recovered whatever com-
posure she had lost, picked up her deck of cards, and
said to herself, "That's the way it goes, folks. The
earth's a very shaky mobile thing, and that's how it
works. Apparently, the mountains around here are
still going up." Later, she would say, "We were
taught all wrong. We were taught that changes on
the face of the earth come in a slow steady march.
But that isn't what happens. The slow steady march
of geologic time is punctuated with catastrophes.
And what we see in the geologic record are the
catastrophes. Look at a graded sandstone and see the
bedding go from fine to coarse. That's a storm. That's
one storm—when the water came up and laid the
coarse material down over the fine. In the rock record, the tranquillity of time is not well represented.
Instead, you have the catastrophes. In the Southwest,
they live from one catastrophe to another, from one
flash flood to the next. The evolution of the world
does not happen a grain at a time. It happens in the
hundred-year storm, the hundred-year flood. Those
things do it all. That earthquake made a catas-
trophist of me."

No one knew where the bears went when they
left the Gallatin Range. When they came back, they
were covered with mud.

·   ·   ·

Catastrophism in another form presented itself
that autumn when Jack Epstein was transferred to

the office of the Geological Survey's Water Resources Division in Alexandria, Louisiana. There was no position for Anita, and she could not have had a job even if one had been open, for it was a rule of the Survey that spouses could not work for the same supervisor. The Alexandria office was small, and included one supervisor. Her nascent geological career was suddenly aborted. She taught physics and chemistry in a Rapides Parish high school. In the summer that followed, she worked for the state government as an interviewer in the unemployment office. She did her geology when and where she could. Driving home from work, she saw people dressed like signal flags hitting golf balls on fake moraines.

Fortunately, her husband was even less interested in the water resources of Louisiana than she was in the unemployment interviews. They decided they needed Ph.D.s to improve their chances of working somewhere else. They enrolled at Ohio State, and in eastern Pennsylvania took up the summer fieldwork that led to their dissertations. They did geologic mapping and biostratigraphy among the ridges of the folded Appalachians—noting the directional trends of the various formations (the strike) and their angles of dip, along a narrow band of deformation from the Schuylkill Gap near Reading to the Delaware Water Gap, and on toward the elbow of the Delaware River where Pennsylvania, New Jersey, and New York conjoin. The most recent ice sheet had reached the Water Gap—where the downcut-

ting Delaware River had sawed a mountain in two—
and had filled the gap, and even overtopped the
mountain, and then had stopped advancing. So the
country of their dissertations was filled with fossil
tundra, with kames and eskers, with periglacial
boulders and the beds of vanished lakes, with erratics from the Adirondacks, with a vast imposition
of terminal moraine. Like the outwash of Brooklyn
and the tills of Indiana, this Pennsylvania countryside helped to give Anita her sophistication in glacial
geology, which was consolidated at Ohio State,
whose Institute of Polar Studies trains specialists in
the field. Glacial evidence was not, however, what
drew her particular attention. The Wisconsinan ice
was modern, in the long roll of time, in much the
way that Edward VII is modern compared with a
hominid skull. The ice melted back out of the Water
Gap seventeen thousand years ago. Anita was more
interested in certain stratigraphic sequences in rock
that protruded through the glacial debris and had
existed for several hundred million years. She would
crush this rock, separate out certain of its components, and under a microscope at fifty to a hundred magnifications study its contained conodonts,
hard fragments of the bodies of unknown marine
creatures—hard as human teeth, and of the same
material. At a hundred magnifications, some of them
looked like wolf jaws, others like shark teeth, arrowheads, bits of serrated lizard spine—not unpleasing
to the eye, with an asymmetrical, objet-trouvé ap-

ary changes in her specimens that some were light
and some were dark. They were white, brown, yellow, tan, and gray. Since they were coming into Columbus from all over the United States, and in fact
the world, she began to notice that in a general way
their colors followed geographical patterns. She
wondered what that might suggest. She looked at
conodonts from Kentucky and Ohio, which were of a
yellow so pale it was almost white. From western
Pennsylvania they were jonquil, from central Pennsylvania brown. The ones she had collected north of
Schuylkill Gap were black. She thought at first there
was something wrong with her samples, but her ad-
viser told her that in all likelihood the blackness was
merely the result of pressures attendant when the
limestone or dolomite was being deformed. He did
not encourage her to make a formal study of the
matter, and she returned to her absorptions with
conodont biostratigraphy. On one of her trips East,
she crossed New York State, collecting dolomite and
limestone all the way. From Lake Erie to the Cat-
skills, New York State is a cake of Devonian rock,
lying flat in a swath sixty miles wide. You can travel
across it chipping off rock of the same approximate
age, and not just any old Devonian samples—for the
Devonian period covers fifty million years—but, say,
limestone and dolomite from the Gedinnian age,
which is eight million years of early Devonian, or
even from the Helderbergian stage of middle Gedinnian time. For as much as a hundred and fifty miles,

[47]

what she could get, which was a map-editing job in Washington. She would have preferred to work on conodonts, but the federal budget at that time covered only one conodont worker, and someone else had the job. Before long, she had become general editor of all geologic mapping taking place east of the Mississippi River. She dealt with hundreds of geologists. There were fifteen hundred in the Survey, and the quality of their work, their capacity for visualizing plunging synclines and recumbent folds, tended to vary. She looked upon some of them as "losers." Such people were sent to what she privately described as "penal quadrangles": the lesser bayous of Louisiana, the Okefenokee Swamp. If they did not know strike from dip, they could go where they would encounter neither. She did not feel pity. Better to be a loser in the United States, she thought, than to be a geological peasant in China. There are four hundred thousand people in the Chinese geological survey. "It's a hell of an outfit," in Anita's words. "If they want to see exposed rock, they don't depend on streambanks and roadcuts, as we do. If an important Chinese geologist wants to see a section of rock, the peasants dig out a mountainside."

She was a map editor for seven years, during all of which she continued her conodont research, almost wholly on her own time. Collecting rock from Maryland and Pennsylvania, she crushed it and "ran the samples" at home. Running samples was not just a matter of pushing slides past the nose of a microscope. After pulverizing the rock and dissolving most of it in acid, she had to sort its remaining components, and this could not be done chemically, so it had to be done physically. It was a problem analogous to the separation of uranium isotopes, which in the early nineteen-forties had brought any number of physicists to a halt. It was also something like sluicing gold, but you could not see the gold.

Anita uses tetrabromoethane, an extremely heavy and extremely toxic fluid that costs a hundred and fifty dollars a gallon. Granite will float in tetrabromoethane. Quartz will float in tetrabromoethane. Conodonts sink without a bubble. Her hands in rubber gloves within a chemical hood, she pours the undissolved rock residue into the tetrabromoethane. The lighter materials float—limestone, dolomite, quartz. Inconveniently, conodonts are not all that sink in tetrabromoethane. Pyrite, among other things, sinks, too. With methylene iodide, a fluid even heavier than tetrabromoethane, she turns the process around. In methylene iodide, the pyrite and whatnot go to the bottom, while the conodonts, among other things, float. Electromagnetically, she further concentrates the conodonts. She can now have a look at them under a microscope, seeing "bizarre shapes that any idiot can recognize," and assign them variously to the Anisian, Ladinian, Cayugan, Osagean, Llandoverian, Ashgillian, or any other among tens of dozens of subdivisions of Cambrian, Ordovician, Silurian, Devonian, Mississippian, Pennsylvanian, Permian, and Triassic time.

[50]

While recording ages, she could not ignore colors, and the question of their possible significance returned to her mind. In the Appalachians generally, formations thickened eastward. The farther east you went, the deeper the rock had once been buried—the greater the heat had once been. Heat appeared to her to have affected the color of the conodonts in the same manner that it affects the color of butter—turning it from yellow to light brown to darker brown to black-and-ruined smoking in the pan. Oh, she thought. You could use those things as thermometers. They might help in mapping metamorphic rock. Metamorphism, the process by which heat and pressure change one kind of rock into another—turn shale into slate, turn granite into gneiss, turn limestone into marble—is divided into grades of intensity. Maybe conodont colors, plotted on a map, could demonstrate the shadings of the grades. At work, she began saying to people, "Show me a conodont and I'll tell you where in the Appalachians it came from." With amazing accuracy she repeatedly passed the test. She imagined that color had been controlled by carbon fixing. In the presence of heat, she thought, the amount of carbon in a given conodont would have remained constant while the amounts of hydrogen and oxygen declined, which is what happens in heated butter. No one seemed to agree with her. One way to test her idea might have been to scan for individual elements with an electron probe, but this was 1967 and electron probes in those days could not pick up light elements like hydrogen

"Yes," he said. "You can. And you can also do it by observing changes in organic materials such as fossil pollen and spores, where they exist."

"How do you do *that*?" she said.

"By looking at color change," he said. "You see, the pollen and spores—"

"Stop!" she said. "Stop right there. They change from pale yellow to brown to black. Am I right?"

"Right," he said. He was matter-of-fact in tone. He was, among other things, an oil geologist, while she was not. Oil companies had been using the colors of fossil pollen and fossil spores to help identify rock formations that had achieved the sorts of temperatures in which oil might form. Land-based plants, with their pollen and spores, had not developed on earth until a hundred and fifty million years after the beginning of the Paleozoic era, however. Nor would they ever be as plentiful and as nearly ubiquitous as marine fossils. Hearing Leonard Harris mention oil companies and their use of color alteration in pollen and spores, Anita realized in the instant that she had —in her words—"reinvented the wheel." And then some. She had not known that pollen and spores were used as geothermometers in the oil business, and now that she knew it she could see at once that conodonts used for the same purpose would have different geographical applications, covering greater ranges of temperature and different segments of time.

"I think I can do the assessments easier and better by using conodonts," she said to Harris. "Conodonts change color, too, and in the same way."

It was his turn to be surprised. "How come I never heard about that?" he said.

She said, "Because no one knows it."

.   .   .

Petroleum—the transmuted fossils of ocean algae—forms when the rock that holds the fossils becomes heated to the temperature of a cup of coffee and remains as warm or warmer for at least a million years. The minimal temperature is about fifty degrees Celsius. At lower temperatures, the algal remains will not turn into oil. At temperatures hotter than a hundred and fifty degrees, any oil or potential oil within the rock is destroyed. ("The stuff is there, throughout the Appalachians. You look at the rocks and you see all this dead oil.") The narrow "petroleum window," as it is called—between fifty degrees and a hundred and fifty degrees—is scarcely a fourteenth part of the full temperature variation of the crust of the earth, a fact that goes a long way toward explaining how the human race could have used up such a large part of the world's petroleum in less than a century. Not only must the marine algae have been buried for adequate time at depths where temperatures hover in the window but once oil has formed it is subject to destruction underground if for one reason or another the temperature of its host rock rises.

Natural gas is to oil as politicians are to states-men. Any organic material whatsoever will form natural gas, and will form it rapidly, at earth-surface temperatures and on up to many hundreds of de-grees. In Anita's words: "You get natural gas as soon as anything drops dead. For oil, the requisites are the organic material and the thermal window. When they look for oil, they don't know what they've got until they drill a hole." In trying to figure out where to drill, geologists have an obvious need for geo-thermometers. Pollen and spores are of considerable use, but only when they have fossilized in certain rocks. Moreover, they are absent altogether from early Paleozoic times, and they are extremely rare in rock from the deep sea.

Leonard Harris asked Anita how many years she had been "sitting on" her discovery about conodonts.

About ten, she told him. The last thing she had wished to do was to keep it secret, but no one had shown much interest. She gave him slides of the New York State east-west series, and told him that a com-parable set could be got together for Pennsylvania, too. Harris went south and traversed the state of Tennessee, collecting carbonate rocks that were close in age, and when Anita ran the conodonts she found the color alterations quite the same as in the north-ern states—dark in the east, pale in the west. Leonard and Anita reported all this to Peter Rose, leader of the Oil & Gas Branch, pointing out that the variations in conodont color could lead to a cheap and rapid technique of finding rock in the petroleum window. Rose said he couldn't understand why no one in the United States had ever thought of this if it was as obvious as all that. Anita told him that for years she had been puzzled by the same question, since the procedure would be one that "any idiot ought to be able to follow, because all you need is to be not color-blind."

At Rose's request, Anita's division of the Geological Survey allowed her to work two days a week on conodonts. Weekends, she worked on them at home. Actual temperature values had not been assigned to the varying colors. She did so in a year of experiments. She began with the palest of conodonts from Kentucky and heated them at varying temperatures until they became canary and golden and amber and chocolate and cordovan, black, and gray. With enough added heat, they would turn white and then clear. At nine hundred degrees Celsius, they disintegrated. By cooking her samples in a great many variations of the ratio of time to temperature, she was able to develop a method of extrapolating laboratory findings onto the scale of geologic time. She concluded that pale-yellow conodonts could remain at about fifty degrees indefinitely without changing color. If they were to remain at sixty to ninety degrees for a million years or more, they would be amber. The earth's thermal gradient varies locally, but generally speaking the temperature of rock increases about one degree Celsius for each hundred feet of depth. A conodont would have to be lodged in rock buried three thousand to six thousand feet in order to experience temperatures of the sort that would turn it amber. At depths of nine thousand to fifteen thousand feet, she discovered, conodonts would turn light brown in roughly ten million years. If they spent ten million years at, say, eighteen thousand feet, they would be dark brown. In comparable amounts of time but at greater and greater depths, they would turn black, gray, opaque, white, clear as crystal. Anita also cooked conodonts in pressure bombs, because it had been suggested to her that the pressures of great tectonism—the big dynamic events in the crust, with mountains building and whole regions being kneaded like dough—might also affect conodont colors. Her experiments convinced her that pressure has little effect on color; heat is what primarily causes it to change.

Of course, plenty of heat is produced by deep burial during major tectonic events. Her conodonts from New Jersey were black and from Kentucky pale essentially because huge disintegrating Eastern mountain ranges had buried the near ones very deep and the far ones scarcely at all. The East is for the most part the wreckage of the ancestral Appalachians, and—as is exemplified in the Devonian rock of New York—the formations are thickest close to where the mountains stood. A continuous sedimentary deposit that is thousands of feet thick in eastern Pennsylvania may be ten feet thick in Ohio.

Where oil was first discovered in western Pennsylvania, it was seeping out of rocks and running in the streams. As a natural lubricant, it is of a character and purity so remarkable that it can virtually be put into a Mercedes without first passing through a refinery. People used to buy it and drink it for their health. Anita looked at conodont samples from rock that surrounded this truly exceptional oil. In the temperature range of eighty to a hundred and twenty degrees, they were in the center of the petroleum window. They were golden brown.

With a year of tests run, with Kodachrome pictures, with graphs and charts of what she called her "wind-tunnel models," she was prepared to tell her story. The Geological Society of America was to meet in Florida in November, 1974, and she arranged to deliver a paper there. "I prepared carefully —I always do—so I wouldn't phumpfer. But the G.S.A. meeting was not momentous. They were academics, and not particularly knowledgeable about exploration techniques." Five months later, scarcely knowing what to anticipate, she went to Dallas and spoke before the American Association of Petroleum Geologists. It was the same show, but this time it was playing in the right house. Requests and invitations poured upon her from oil companies wherever they might be, and from geological societies situated in oil centers like Calgary and Tulsa. "It filled a big hole in their technology," Anita has said, recalling those days. "They have to be able to assess the thermal level of deposits, and this was a simple way to do it."

Anita is now a conodont specialist for the United States Geological Survey, full time. She lives in Maryland. Her home is an island in flower beds and lawn. On weekday mornings, she gets up at five-thirty and rides to work in Washington on a Trailways bus. Her laboratory is in the Smithsonian, where she is largely on her own in shaping her research. Oil companies have continued to beat the path to her door, as have oil geologists from every continent but Antarctica, including large delegations from the Chinese geological survey. While oil prospectors are using brown and yellow conodonts to guide them to the thermal window, mineral prospectors are using white ones in the search for copper, iron, silver, and gold. White conodonts and clear conodonts, products of the highest temperatures, suggest the remains of thermal hot spots, thermal aureoles, ancient hydrothermal springs—places where metallic minerals would have come up in solution to be precipitated out into veins.

Soon after her discovery, universities began calling her, and, ultimately, the American Association for the Advancement of Science. She was pleased to appear at places like Princeton, pleased to be given an opportunity to demonstrate what could be learned elsewhere. Women students were in her audience now. In the late nineteen-seventies, she and her colleagues published a succession of scientific papers whose title pages perforce encapsulated not only their professional endeavors but something of their private lives. The "senior author" of a scientific publication is the person whose name is listed first and whose work has been of primary importance to the project, while other authors are listed more or less in diminishing order, like the ingredients on a can of stew. The benchmark paper came in 1977. Entitled "Conodont Color Alteration—an Index to Organic Metamorphism," it was "by Anita G. Epstein, Jack B. Epstein, and Leonard D. Harris." Then, in 1978, came "Oil and Gas Data from Paleozoic Rocks in the Appalachian Basin: Maps for Assessing Hydrocarbon Potential and Thermal Maturity (Conodont Color Alteration Isograds and Overburden Isopachs)"—virtually an oil-prospecting kit, a highly specialized atlas—"by Anita G. Harris, Leonard D. Harris, and Jack B. Epstein." And scarcely a year after that appeared a summary document called "Conodont Color Alteration, an Organo-Mineral Metamorphic Index, and Its Application to Appalachian Basin Geology"—"by Anita G. Harris."

largely by the woodlots, hedgerows, and striped fields of a broad terrain as much as seven hundred feet lower than the spot on which we stood and of such breathtaking proportions and fetching appearance that it could be mentioned in a sentence with the Shenandoah Valley. The picture of New Jersey that most people hold in their minds does not include a Shenandoah Valley. Nevertheless, this New Jersey Appalachian landscape not only looked like the Shenandoah, it actually was the Shenandoah, in the sense that it was a fragment of a valley that runs south from New Jersey to Alabama and north from New Jersey into Canada—a single valley, one continuous geology, known to science as the Great Valley of the Appalachians and to local peoples here and there as Champlain, Shenandoah, Tennessee Valley, but in New Jersey by no special name. This integral, elongate, predominantly carbonate valley disappears and reappears through the far Northeast, until in pieces it presents itself in Newfoundland and then dives under the sea. Its marbles are minable in Vermont, in Tennessee. It was the route of armies—the avenue to Antietam, the site of Chickamauga, Saratoga, Ticonderoga. It stands in the morning shadow of the Annieopsquotch Mountains, of the Green Mountains, of the New Jersey Highlands, of the Berkshire, Catoctin, and Great Smoky Mountains, which are fraternal in structure and composition and are all of Precambrian age. The lookoff where we stood was a part of that Appalachian complex. It was crystalline rock above a thousand million years old—and the rock in the valley was younger, and in the Kittatinny younger still. (Geologists avoid the word "billion" because in one part and another of the English-speaking world the quantity it refers to differs by three orders of magnitude. A billion in Great Britain is a million million.) We were looking from the New Jersey Highlands into another segment of the cordillera—the beginnings of the physiographic province of the Valley and Ridge, the folded-and-faulted deformed Appalachians, the long ropy ridges of the eastern sinuous welt, which Edmund Wilson had once written off as "fairly unimportant creases in the earth covered with trees."

"Geology repeats itself," Anita remarked, and she went on to say that anyone who could understand the view before us would have come to understand in a general way the Appalachians as a whole —that what we were looking at was the fragmental evidence and low remains of alpine massifs immeasurably high and wide, massifs which for the most part had stood behind us to the east, and were now largely disintegrated and recycled into younger rock that is tens of thousands of feet deep and wedges out to the west in ever-diminishing quantity until what covers Ohio is a thin veneer.

The appearance of a country is the effect primarily of water, running off the landscape, cutting out valleys, dozing wantonly as glacier ice. The sculpturing is external. But it is influenced and can even be controlled by the rock within: by the relative strength, not to mention the solubility, of successive strata, and by the folds and faults—the structure—that the rock has been given. Figuring out the Appalachians was Problem 1 in American geology, and a difficult place to begin, for it was scarcely a matter of layer-cake legibility, like the time scale in the walls of the Grand Canyon. It was a compressed, chaotic, ropy enigma four thousand kilometres from end to apparent end, full of over-turned strata and recycled rock, of steep faults and horizontal thrust sheets, of folds so tight that what had once stretched twenty miles might now fit into five. The country seemed to consist of parallel mean-dering belts—the Piedmont, the Precambrian high-lands, the Great Valley, the folded-and-faulted de-formed mountains, the Allegheny Plateau. It was high and resistant, low and vulnerable. (I have heard the Shenandoah described, if not dismissed, as "a strip of weak rock.") The early Appalachian ge-ologists, in their horse-drawn buggies, their suits and ties, developed a sense of physiography that tuned them to the land, and when they saw long sugarloaf hills they had learned to suspect that there was dolomite within, and when they looked up at cox-comb ridges they felt the presence of Cambrian sandstones, and of Cambrian shales in the valleys beyond. The higher, harder ridges would be thick, Silurian quartzites, more often than not, while flour-ishing green lowlands with protruding ribs of rock

would owe their shape and their fertility to limestones assembled in Ordovician seas. There were
knolls in the valleys. Inside the knolls were shales.
Shale breaks up easily but will not dissolve like limestone, so the shales became blisters in the limestone
valleys. Of the two carbonate rocks, limestone is a
good deal more soluble than dolomite, and that was
why dolomite would retain itself in sugarloaves
above the limestone valleys. Once the early geologists
had developed this sense of the substrate, they shook
the reins and moved with dispatch, filling in the first
American geologic maps with a general accuracy
that is impressive still.

Identifying what is there scarcely describes
what happened to put it there, however. The history
of the earth may be written in rock, but history is not
coherent on a geologic map, which shows a region's
uppermost formations in present time, while indicat-
ing little of what lies farther down and less of what is
gone from above. At a given place—a given latitude
and longitude—the appearance of the world will
have changed too often to be recorded in a single
picture, will have been, say, at one time below fresh
water, at another under brine, will have been moun-
tainous country, a quiet plain, equatorial desert, an
arctic coast, a coal swamp, and a river delta, all in
one Zip Code. These scenes are discernible in, among
other things, the sedimentary characteristics of rock,
in its chemical composition, magnetic components,
interior color, hardness, fossils, and igneous, meta-

morphic, or depositional age. But as parts of the historical narrative these items of evidence are just phrases and clauses, often wildly disjunct. They are like odd pieces from innumerable jigsaw puzzles. The rock column—a vertical representation of the crust at some point on the earth—holds a great deal of inferable history, too. But rock columns are generalized; they are atremble with hiatuses; and they depend in large part on well borings, which are shallow, and on seismic studies, which are new, and far between. To this day, in other words, there remains in geology plenty of room for the creative imagination. All the more amazing is the extent to which the early geologists, who travelled the Appalachians in the eighteen-twenties and thirties, not only catalogued the evident rock but also worked out stratigraphic relationships among various formations and began to see composite structure. Starting close up, with this rock type, that mountain, this formation, that valley—with what they could see and know— they gradually began to form tentative regional pictures. Piece by piece over the next century and a half, they and their successors would put together logically sequential narratives presenting the comprehensive history of the mountain belt. As new evidence and insight came along, old logic sometimes fell into discard. When plate tectonics arrived, its revelations were embraced or accommodated but by no means universally accepted. The Appalachians, meanwhile, continued slowly to waste away. The debate about their origins did not.

Observing the valley scene, the gapped and distant ridgeline, Anita said that mountains in this region had come up and been worn down not once but a number of times: the Appalachians were the result of a series of pulses of mountain-building, the last three of which had been spaced across two hundred and fifty million years and were known as the Taconic Orogeny, the Acadian Orogeny, and the Alleghenian Orogeny. The first stirrings of the Taconic Orogeny began five hundred million years ago. After the mountains it lifted had been pretty much eroded away, their stubs and their detritus, much of which had turned into sedimentary rock, became involved in the Acadian Orogeny; and when the Acadian Orogeny was long gone by, its mountain stubs and lithified debris were caught up in the Alleghenian Orogeny, which drove into the sky still another massif, the ruins of which lay all about us now. In such manner had each of the orogenies of the Appalachians cannibalized the products of previous pulses, and now we were left with this old mountain range, by weather almost wholly destroyed, but nonetheless containing in a traceable and unarguable way the rock of its ancestral mountains. She said the Delaware Water Gap, with its hard quartzites, represented action from the heart of the story, debris from the Taconic Orogeny: boulders, pebbles, sands, and silts carried down from bald mountains by the rapids of big braided rivers—a runoff unimpeded by vegetation, when not so much as one green leaf existed in the terrestrial world.

Long before the Taconic mountain-building pulse was felt, the scene was very different. A subdued continent, consisting of what is now the basement rock of North America, stood low with quiet streams, collecting on its margins clean accumulations of sand. One can infer the flat landscape, the slow rivers, the white beaches, in the rock that remains from those Cambrian sands. Sea level, never constant, moved generally upward all through Cambrian time. The water advanced upon the continent at an average rate of ten miles every million years, spreading across the craton successive coastal sands. Potsdam sandstone. Antietam sandstone. Waynesboro sandstone. Eau Claire sandstone. There were seventy million years in the long tectonic quiet of Cambrian time, 570 to 500 million years before the present. By the end of the Cambrian and the beginning of the Ordovician, the ocean had spread its great bays upon the continents to an extent that has not been equalled in five hundred million years, with the possible exception of the highest Cretaceous seas. No one knows why. There is a fixed amount of water in the world. It can rain and run, evaporate, freeze, sit in deep cold pools on abyssal plains, but it cannot leave the earth. When large amounts of it collect as ice upon the continents, the level of the sea drastically goes down. In much of Cambro-Ordovician time, glaciation was absent from the world, and almost all water was in a fluid state. But that alone will not explain the signal height of the band of the Lenape. They came into the region toward the dawn of Holocene time and lost claim to it in the beginnings of the Age of Washington. Like index fossils, they now represent this distinct historical stratum. Their home and prime hunting ground was the Minisink—over the mountain, beside the river, the country upstream from the gap. The name Delaware meant nothing to them. It belonged to a family of English peers. The Lenape named the river for themselves. I knew some of this from my grade-school days, not many miles away. The Minisink is a world of corn shocks and islands and valley mists, of trout streams and bears, today. Especially in New Jersey, it has not been mistreated, and, with respect to the epoch of the Minsi, geologically it is the same. The Indians of the Minisink were good geologists. Their trails ran great distances, not only to other hunting parks and shell-mounded beach camps but also to their quarries. They set up camps at the quarries. They cooked in vessels made of soapstone, which they cut from the ground in what is now London Britain Township, Chester County, Pennsylvania. They made adzes of granite, basalt, argillite, even siltstone, from sources closer to home. They went to Berks County, in Pennsylvania, for gray chalcedony and brown jasper. They used glacial-erratic hornfels. They made arrowheads and spear points of Deepkill flint. They made drills and scrapers of Onondaga chert. Flint, chert, and jasper are daughters of chalcedony, which in turn is a variety own magnetite, turning green from trace copper. Its appearance can be deceiving. Geologists are slow to identify exposures they have not seen before. They don't just cruise around ticking off names at distances that would impress a hunter. They go up to outcrops, hit them with hammers, and look at the rock through ten-power lenses. If the possibilities include the carbonates, they try a few drops of hydrochloric acid. Limestone with hydrochloric acid on it immediately forms a head, like beer. Dolomite is less responsive to acid. With her sledgehammer, Anita took many pounds of roadcut, and not without effort. Again and again, she really had to slam the wall. Looking at the fresh surface of a piece she removed, she said she'd give odds it was dolomite. It was not responsive to acid. She scraped it with a knife and made powder. Acid on the powder foamed. "This dolomite is clean enough to produce beautiful white marble if it were heated up and recrystallized," she said. "When it became involved in the mountain-building, if it had got up to five hundred degrees it would have turned into marble, like the Dolomites, in Italy. There is not a lot of dolomite in the Dolomites. Most of the rock there is marble." She pointed in the roadcut to the domal structures of algal stromatolites—fossil colonies of microorganisms that had lived in the Cambrian seas. "You know the water was shallow, because those things grew only near the light," she said. "You can see there was no mud around. The rock is so clean. And you know the water was warm, because you do not get massive carbonate deposition in cold water. The colder the water, the more soluble carbonates are. So you look at this roadcut and you know you are looking into a clear, shallow, tropical sea."

With dry land adrift and the earth prone to rolling, that Cambrian sea and New Jersey below it would have been about twenty degrees from the equator—the present latitude of Yucatán, where snorkelers kick along in transparent waters looking through their masks at limestones to be. The Yucatán peninsula is almost all carbonate and grew in its own sea. As did Florida. Under the shallow waters of the Bahamas are wave-washed carbonate dunes, their latitude between twenty and twenty-six degrees. At the end of Cambrian time, the equator crossed what is now the North American continent in a direction that has become north-south. The equator came in through the Big Bend country in Texas and ran up through the Oklahoma panhandle, Nebraska, and the Dakotas. If in late Cambrian time you had followed the present route of Interstate 80, you would have crossed the equator near Kearney, Nebraska. In New Jersey, you would have been in water scarcely above your hips, wading among algal mounds and grazing gastropods. You could have waded to the equator. West of Chicago and through most of Illinois, you would have been wading on clean sand, the quiet margin of the Canadian craton, which remained above the sea. The limy bottom apparently resumed in Iowa and went on into eastern Nebraska, and then, more or less at Kearney, you would have moved up onto a blistering-hot equatorial beach and into low terrain, subdued hills, rock that had been there a thousand million years. It was barren to a vengeance with a hint of life, possibly a hint of life—rocks stained green, stained red by algae. Wyoming. Past Laramie, you would have come to a west-facing beach and, after it, tidal mudflats all the way to Utah. The waters of the shelf would now begin to deepen. A hundred miles into Nevada was the continental slope and beyond it the blue ocean.

If you had turned around and gone back after fifty million years—well into Ordovician time, say four hundred and sixty million years ago—the shelf edge would still have been near Elko, Nevada, and the gradually rising clean-lime seafloor would have reached at least to Salt Lake City. Across Wyoming, there may have been low dry land or possibly continuing sea. The evidence has almost wholly worn away, but there is one clue. In southeastern Wyoming, a diamond pipe came up about a hundred million years ago, and, in the tumult that followed the explosion, marine limestone of late Ordovician age fell into the kimberlite and was preserved. In western Nebraska, you would have crossed dry and barren Precambrian terrain and by Lincoln have reached another sea. Iowa, Illinois, Indiana. The water was clear, the bottom uneven—many shallows and deeps upon the craton. In Ohio, the sea would

have begun to cloud, increasingly so as you moved
on east, silts slowly falling onto the lime. In Pennsylvania, as you approached the site of the future
Delaware Water Gap, the bottom would have fallen
away below you, and where it had earlier been close
to the surface it would now be many tens of fathoms
down.

"The carbonate platform collapsed," Anita said.
"The continental shelf went down and formed a big
depression. Sediments poured in." Much in the way
that a sheet of paper bends downward if you move
its two ends toward each other in your hands, the
limestones and dolomites and the basement rock beneath them had subsided, forming a trough, which
rapidly filled with dark mud. The mud became
shale, and when the shale was drawn into the heat
and pressure of the making of mountains its minerals realigned themselves and it turned into slate.
We moved on west a couple of miles and stopped
at a roadcut of ebony slate. Anita said, "Twelve
thousand feet of this black mud was deposited
in twelve million years. That's a big pile of rock."
The formation was called Martinsburg. It had been
folded and cleaved in orogenic violence following
its deposition in the sea. As a result, it resembled
stacks of black folios, each of a thousand leaves. Just
to tap at such rock and remove a piece of it is to
create something so beautiful in its curving shape
and tiered laminations that it would surely be attractive to a bonsai gardener's eye. It seems a proper
setting for a six-inch tree. I put a few pieces in the

car, as I am wont to do when I see some Martinsburg. Across the Delaware, in Pennsylvania, the formation presents itself in large sections that are without joints and veins, the minerals line up finely in dense flat sheets, and the foliation planes are so extensive and straight that slabs of great size can be sawed from the earth. The rock there is described as "blue-gray true unfading slate." It is strong but "soft," and will accept a polishing that makes it smoother than glass. From Memphis to St. Joe, from Joplin to River City, there is scarcely a hustler in the history of pool who has not racked up his runs over Martinsburg slate. For anybody alive who still hears corruption in the click of pocket billiards, it is worth a moment of reflection that not only did all those pool tables accumulate on the ocean floor as Ordovician guck but so did the blackboards in the schools of all America.

The accumulation of the Martinsburg—the collapsing platform, the inpouring sediments—was the first great sign of a gathering storm. Geological revolution, crustal deformation, tectonic upheaval would follow. Waves of mountains would rise. Martinsburg time in earth history is analogous to the moment in human history when Henry Hudson, of the Dutch East India Company, sailed into the bay of the Lenape River.

Completing the crossing of the Great Valley of the Appalachians, Anita and I passed more limestones, more slate. Their original bedding planes, where we could discern them, were variously atilt, vertical, and overturned, so intricate had the formations become in the thrusting and folding of the long-gone primal massifs. The road came to the river, turned north to run beside it, and presented a full view of the break in the Kittatinny ridge, still far enough away to be comprehended in context but close enough to be seen as the phenomenon it is: a mountain severed, its folds and strata and cliffs symmetrical, thirteen hundred feet of rock in close fraternal image from the skyline to the boulders of a blue-and-white river. Small wonder that painters of the Hudson River School had come to the Delaware to do their best work. George Inness painted the Delaware Water Gap many times, and he chose this perspective—downriver about four miles—more than any other. I have often thought of those canvases—with their Durham boats on the water and cows in the meadows and chuffing locomotives on the Pennsylvania side—in the light of Anita's comment that you would understand a great deal of the history of the eastern continent if you understood all that had made possible one such picture. She was suggesting, it seemed to me, a sense of total composition—not merely one surface composition visible to the eye but a whole series of preceding compositions which in the later one fragmentarily endure and are incorporated into its substance—with materials of vastly differing age drawn together in a single scene, a composite canvas not only from the Hudson River

School but including everything else that had been a part of the zones of time represented by the boats, gravels, steeples, cows, trains, talus, cutbanks, and kames, below a mountain broken open by a river half its age.

The mountain touched the Martinsburg, and its rock was the younger by at least ten million years. Kittatinny Mountain is largely quartzite, the primary component of the hubs of Hell. In the post-tectonic, profoundly eroded East, quartzite has tended to stand up high. The Martinsburg is soft, and is therefore valley. There is nothing but time between the two. Where the formations meet, a touch of a finger will cover both the beginning and the end of the ten million years, which are dated at about 440 and 430 million years before the present—from latest Ordovician time to a point in the early Silurian. During that time, something apparently lifted the Martinsburg out of its depositional pit and held it above sea level until weathers wore it low enough to be ready to accept whatever might spread over it from higher ground. The quartzite—as sand—spread over it, coming down from Taconic mountains. The sand became sandstone. Upward of fifty million years later, the sand grains fused and turned into quartzite in the heat and the crush of new rising mountains, or possibly a hundred million years after that, in the heat and the crush of more mountains. The Delaware River at that time was not even a cloud in the sky. Rivers of greater size were flowing the other way, must have widened the Water Gap. It gouged out the riverbed and left there afterward two hundred feet of gravel. Indians were in the Minisink when the vegetation was tundra. Ten thousand years ago, when the vegetation changed from tundra to forest, Indians in the Minisink experienced the change. The styles in which they fractured their flint—their jasper, chert, chalcedony—can be correlated to Anatolian, Sumerian, Mosaic, and Byzantine time. Henry Hudson arrived in the New World about four hundred years before the present. He was followed by Dutch traders, Dutch colonists, Dutch miners. They discovered ore-grade copper in the Minisink, or thought they did. Part fact, part folklore, it is a tradition of the region that a man named Hendrik Van Allen assessed Kittatinny Mountain and decided it was half copper. The Dutch crown ordered him to establish a mine, and to build a road on which the ore could be removed. The road ran up the Minisink and through level country to the Hudson River at Esopus Creek (Kingston, New York). A hundred miles long, it was the first constructed highway in the New World to cover so much distance. It covers it still, and is in many places scarcely changed. When Van Allen was not busy supervising the road builders, he carried on an élite flirtational minuet with the daughter of a Lenape chief. The chief was Wissinoming, his daughter Winona. One day, Van Allen went alone to hunt in the woods near the river islands of the Minisink, and he discharged his piece in the direction of a squirrel. The creature scurried through the branches of trees. Van Allen shot again. The creature scurried through the branches of other trees. Van Allen reloaded, stalked the little bugger, and, pointing his rifle upward, sighted with exceptional care. He fired. The squirrel fell to the ground. Van Allen retrieved it, and found an arrow through its heart. By the edge of the river, Winona threw him a smile from her red canoe. They fell in love. In the Minisink, there was no copper worth mentioning. Van Allen didn't care. Winona rewrote the country for him, told him the traditions of the river, told him the story of the Endless Mountain. In the words of Winona's legend as it was eventually set down, "she spoke of the old tradition of this beautiful valley having once been a deep sea of water, and the bursting asunder of the mountains at the will of the Great Spirit, to uncover for the home of her people the vale of the Minisink." In 1664, Peter Stuyvesant, without a shot, surrendered New Amsterdam and all that went with it to naval representatives of Charles, King of England. Word was sent to Hendrik Van Allen to close his mines and go home. It was not in him to take an Indian wife to Europe. He explained these matters to Winona in a scene played out on the cliffs high above the Water Gap. She jumped to her death and he followed.

. . .

On foot at the base of the cliffs—in the gusts and shattering noise of the big tractor-trailers passing almost close enough to touch—we walked the narrow space between a concrete guard wall and the rock. Like the river, we were moving through the mountain, but in the reverse direction. Between the mirroring faces of rock, rising thirteen hundred feet above the water, the gap was so narrow that the interstate had been squeezed in without a shoulder. There was a parking lot nearby, where we had left the car—a Delaware Water Gap National Recreation Area parking lot, conveniently placed so that the citizen-traveller could see at point-blank range this celebrated natural passage through a mountain wall, never mind that it was now so full of interstate, so full of railroad track and other roadways that it suggested a convergence of tubes leading to a patient in Intensive Care. We saw painted on a storm sewer a white blaze of the Appalachian Trail, which came down from the mountain in New Jersey, crossed the river on the interstate, and returned to the ridgetop on the Pennsylvania side. There were local names for the sides of the gap in the mountain. The Pennsylvania side was Mt. Minsi, the New Jersey side Mt. Tammany. The rock of the cliffs above us was cleanly bedded, stratified, and had been not only deposited but also deformed in the course of the eastern orogenies. Regionally, it had been pushed together like cloth on a table. The particular fragment of the particular fold that erosion had left as the sustaining rock of the mountain happened to be dipping to the northwest at an angle of some forty-five degrees. As we walked in that general direction, each upended layer was somewhat younger than the last, and each, in the evidence it held, did not so much suggest as record progressive changes in Silurian worlds. "The dip always points upsection, always points toward younger rocks," Anita said. "You learn that the first day in Geology I."

"Do you ever get tired of teaching ignoramuses?" I asked her.

She said, "I haven't worked on this level since I don't know when."

Near the road and the river, at the beginning of the outcrop, great boulders of talus had obscured the contact between the mountain quartzite and the underlying slate. To move on through the gap, traversing the interior of the mountain, was to walk from early to late Silurian time, to examine an assembly of rock that had formed between 430 and 400 million years before the present. The first and oldest quartzite was conglomeratic. Its ingredients had lithified as pebbles and sand. Shouting to be heard, Anita said, "In those pebbles you can see a mountain storm. You can see the pebbles coming into a sandbar in a braided river. There is very little mud in this rock. The streams had a high enough gradient to be running fast and to carry the mud away. These sands and pebbles were coming off a mountain range, and it was young and high."

A braided river carries such an enormous burden of sand and gravel that it does not meander through its valley like most streams, making cutbanks to one side and point bars opposite. Instead, it runs in braided channels through its own broad bed. Looking at those Silurian conglomerates, I could all but hear the big braided rivers I had seen coming down from the Alaska Range, with gravels a mile wide, caribou and bears on the gravel, and channels flowing in silver plaits. If those rivers testify, as they do, to the erosional disassembling of raw young mountains, then so did the rock before us, with its clean river gravels preserved in river sand. "Geology repeats itself," Anita said, and we moved along, touching, picking at the rock. She pointed out the horse-belly curves of channel-fill deposits, and the fact that none was deeper than five feet—a result of the braiding and the shifting of the channels. Evidently, the calm earth and quiet seas that were described by the older rock we had collected up the road had been utterly revolutionized in the event that built the ancient mountains, which, bald as the djebels of Arabia, had stood to the east and shed the sand and gravel this way. In the ripple marks, the crossbedding, the manner in which the sands had come to rest, Anita could see the westerly direction of the braided-river currents more than four hundred million years ago.

Three hundred years ago, William Penn arrived in this country and decided almost at once that the

Lenape were Jews. "Their eye is little and black, not unlike a straight-look't Jew," he wrote home. "I am ready to believe them of the Jewish race. . . . A man would think himself in Dukes-place or Berry-street in London, when he seeth them." They were "generally tall, straight, well-built" people "of singular proportion." They greased themselves with clarified bear fat. Penn studied their language—the better to know them, the better to work out his treaties. "Their language is lofty, yet narrow, but like the Hebrew. . . . One word serveth in the place of three. . . . I must say that I know not a language spoken in Europe that hath words of more sweetness or greatness, in accent or emphasis, than theirs." Penn heard "grandeur" in their tribal proper names. He listed them: Tammany, Poquessin, Rancocas, Shakamaxon. He could have added Wyomissing, Wissinoming, Wyoming. He made treaties with the Lenape under the elms of Shakamaxon. Tammany was present. He was to become the most renowned chief in the history of the tribe. Many years after his death, American whites in Eastern cities formed societies in his name, and called him St. Tammany, the nation's patron saint. Penn's fondness for the Lenape was the product of his admiration. Getting along with the Lenape was not difficult. They were accommodating, intelligent, and peaceful. The Indians revered Penn as well. He kept his promises, paid his way, and was fair.

Under the elms of Shakamaxon, the pledge was

made that Pennsylvania and the Lenape would be friends "as long as the sun will shine and the rivers flow with water." Penn outlined his needs for land. It was agreed that he should have some country west of the Lenape River. The tracts were to be defined by the distance a man could walk in a prescribed time— typically one day, or two—at an easygoing pace, stopping for lunch, for the odd smoke, as was the Lenape manner. In camaraderie, the Penn party and the Indians gave it up somewhere in Bucks County. Penn went home to England. He died in 1718.

About fifteen years later, Penn's son Thomas, a businessman who had a lawyer's grasp of grasping, appeared from England with a copy of a deed he said his father had transacted, extending his lands to the north by a day and a half's walk. He made it known to a new generation of Lenape, who had never heard of it, and demanded that they acquiesce in the completion of—as it came to be called—the Walking Purchase. With his brothers, John and Richard, he advertised for participants. He offered five hundred acres of land for the fleetest feet in Pennsylvania. In effect, he hired three marathon runners. When the day came—September 19, 1737— the Lenape complained. They could not keep up. But they followed. Their forebears had made a bargain. The white men "walked" sixty-five miles, well into the Poconos. Even so blatant an affront might in time have been accepted by the compliant tribe. But flood waters over mud cracks that had baked in the sun. From a layer of conglomerate, Anita removed a pebble with the pick end of her rock hammer. "Milky quartz," she said. "Bull quartz. We saw this rock back up the road in the Precambrian highlands. When the Taconic Orogeny came, it lifted the older rock, and erosion turned it into pebbles and sand, which is what is here in this conglomerate. It's an example of how the whole Appalachian system continually fed upon itself. These are Precambrian pebbles, in Silurian rock. You'll see Silurian pebbles in Devonian rock, Devonian pebbles in Mississippian rock. Geology repeats itself." Now and again, we came to small numbers that had been painted long ago on the outcrops. Anita said she had painted the numbers when she and Jack Epstein were working on the geology of the Water Gap. She said, "I'd hate to tell you how many months I've spent here measuring every foot of rock." Among the quartzites were occasional bands not only of sandstone but of shale. The shales were muds that had settled in a matter of days or hours and had filled in the lovely periodicity of the underlying ripples in the ancient river sand. For each picture before us in the rock, there was a corresponding picture in her mind: scenes of the early Silurian barren ground, scenes of the rivers miles wide, and, over all, a series of pictures of the big Taconic mountains to the east gradually losing their competition with erosion in the wash of Silurian rain—a general rounding down of things, with river gradients declining. There were pictures of subsiding country, pictures of rising seas. She found shale that had been the mud of an estuary, and fossil shellfish, fossil jellyfish, which had lived in the estuary. In thin dark flakes nearby she saw "a little black lagoon behind a beach." And in a massive layer of clear white lithified sand she saw the beach. "You don't see sand that light except in beaches," she said. "That is beach sand. You would have looked westward over the sea."

To travel then along the present route of Interstate 80, you would have been in need of a seaworthy shallow-draft boat. The journey could have started in mountain rapids, for the future site of the George Washington Bridge was under thousands of feet of rock. Down the huge fans of boulders and gravel that leaned against the mountains, the west-running rivers raced toward the epicontinental sea. They projected their alluvium into the water and spread it so extensively that up and down the long flanks of the Taconic sierra the alluvium coalesced, gradually building westward as an enormous collection of sediment—a deltaic complex. At the future site of the Water Gap, you would have shoved off the white beach and set a westerly course across the sea—looking back from time to time up the V-shaped creases of steep mountain valleys. That was the world in which the older rock of the Water Gap had been forming—the braided-river conglomerates, the estuary mud, the beach sand. In the Holocene epoch, the

Andes would look like that, with immense fans of gravel coming off their eastern slopes—the essential difference being vegetation, of which there was virtually none in the early Silurian. The sea was shallow, with a sandy bottom, in Pennsylvania. The equator had shifted some and was running in the direction that is now northeast-southwest, through Minneapolis and Denver. There were muds of dark lime in the seafloors of Ohio, and from Indiana westward there were white-lime sands only a few feet under clear water.

If you had turned around and come back thirty million years later, in all likelihood you would still have been riding the sparkling waves of the limestone-platform sea, but its extent in the late Silurian is not well reported. Most of the rock is gone. There are widely scattered clues. Among the marine limestones that fell into the diamond pipe near Laramie, some are late Silurian in age. From Wyoming toward the east, there seems to have existed a vastly extrapolated sea. The extrapolation stops in Chicago. You would have come upon a huge coral reef, which is still there, which grew in Silurian time, and did not grow in a desert. It was a wave-washed atoll then, a Kwajalein, an Eniwetok, and in time it would become sugary blue dolomite packed with Silurian shells. After standing there almost four hundred million years, the dolomite would be quarried to become for many miles the concrete surface of Interstate 80 and to become as well the foundations of most of the tall buildings that now proclaim Chicago, as the atoll did in Silurian time. Interstate 80 actually crosses the atoll on bridges above the quarry, which is as close an approximation to the Grand Canyon as Chicago is likely to see, and is arguably its foremost attraction. Beyond the atoll, you would have come to other atolls and hypersaline seas. When water is about three times as salty as the ocean, gypsum will crystallize out. Sticking up from the bottom in central Ohio were dagger-length blades of gypsum crystal. You would have been bucking hot tropical trade winds then, blowing toward the equator, evaporating the knee-deep sea. East of Youngstown, red muds clouded the water—muds coming off the approaching shore. The beach was in central Pennsylvania now, near the future site of Bloomsburg, near the forks of the Susquehanna. The great sedimentary wedge of the delta complex had grown a hundred miles. The Taconic mountains were of humble size. The steep braided rivers were gone, their wild conglomerates buried under meandering mudbanked streams moving serenely through a low and quiet country—a rose-and-burgundy country. There were green plants in the red earth, for the first time ever.

Walking forward through time and past the tilted strata of the Water Gap, we had come to sandstones, siltstones, and shales, in various hues of burgundy and rose. In the irregular laminations of the rock—in its worm burrows, ripples, and crossbeds—

Anita saw and described tidal channels, tidal flats, a river coming into an estuary, a barrier bar, a littoral sea. She saw the delta, spread out low and red, the Taconic mountains reduced to hills. We had left behind us the rough conglomerates and hard gray quartzites that had come off the Taconic mountains when they were high—the formation, known as Shawangunk, that forms the mural cliffs above the Delaware River. (The quartzites are paradisal to rappeling climbers, who refer in their vernacular to "the gap rap," a choice part of "the Gunks.") And now, half a mile up the highway and twenty million years up the time scale, we were looking at the younger of the two formations of which Kittatinny Mountain is locally composed. Generally red, the rock is named for Bloomsburg, outer reach of the deltaic plain in late Silurian time, four hundred million years before the present.

Less than two hundred years before the present, when the United States was twenty-four years old, the first wagon road was achieved through the Water Gap. The dark narrow passage in rattlesnake-defended rock had seemed formidable to Colonial people, and the Water Gap had not served them as a transportational gateway but had been left aloof, mysterious, frightening, and natural.

In the hundred feet or so of transition rocks between the gray Shawangunk and the red Bloomsburg, we had seen the Silurian picture change from sea and seashore to a low alluviated coastal plain;

and if we had a microscope, Anita said, we would see a few fish scales in the Bloomsburg river sands—from fish that looked like pancake spatulas, with eyes in the front corners.

In 1820, the Water Gap was discovered by tourists. They were Philadelphians with names like Binney.

Breaking away some red sandstone, Anita remarked that it was telling a story of cut-and-fill—the classic story of a meandering stream. The stream cuts on one side while it fills in on the other. Where bits and hunks of mudstone were included in the sandstone, the stream had cut into a bank so vigorously that it undermined the muddy soil above and caused it to fall. Meanwhile, from the opposite bank —from the inside of the bend in the river—a point bar had been building outward, protruding into the channel, and the point bar was preserved in clean sandstones, where curvilinear layers, the crossbeds, seemed to have been woven of rushes.

Confronted with a mountain sawed in half, a traveller would naturally speculate about how that might have happened—as had the Indians before, when they supposedly concluded that the Minisink had once been "a deep sea of water." Samuel Preston agreed with the Indians. In 1828, in a letter to *Hazard's Register of Pennsylvania*, Preston referred to the Water Gap as "the greatest natural curiosity in any part of the State." He went on to hypothesize that "from the appearance of so much alluvial or made land above the mountain, there must, in some former period of the world, have been a great dam against the mountain that formed all the settlements called Minisink into a lake, which extended and backed the water at least fifty miles." And therefore, he worked out, "from the water-made land, and distance that it appears to have backed over the falls in the river, the height must, on a moderate calculation, have been between one hundred and fifty and two hundred feet, which would have formed a cataract, in proportion to the quantity of water, similar to Niagara." Preston was a tourist, not a geologist. The first volume of Charles Lyell's "Principles of Geology," the textbook that most adroitly explained the new science to people of the nineteenth century, would not be published for another two years, let alone cross the sea from London. All the more remarkable was Preston's Hypothesis. Like many an accomplished geologist who would follow, Preston made excellent sense even if he was wrong. Withal, he had the courage of his geology. "If any persons think my hypothesis erroneous," he concluded, "they may go and examine for themselves. . . . The Water Gap will not run away."

While sediments accumulated slowly in the easygoing lowlands of the late Silurian world, iron in the rock was oxidized, and therefore the rock turned red. Alternatively, it could have been red in the first place, if it weathered from a red rock source. There were dark-hued muds and light silts in the outcrop, settled from Silurian floods. There were balls and pillows, climbing ripples, flow rolls, and mini-dunes—multihued structures in the river sands. Maroon. Damask. Carmine. Rouge.

Artists were the Delaware Water Gap's most effective discoverers. Inadvertently, they publicized it. They almost literally put it on the map. Arrested by the symmetries of this geomorphological phenomenon, they sketched, painted, and engraved it. The earliest dated work is the Strickland Aquatint, 1830, with a long and narrow flat-bottomed Durham boat in the foreground on the river, four crewmen standing at their oars, a steersman (also standing) in the stern, and in the background the wildwoods rising up the mountain with its deep, improbable incision.

Cutting and filling, a stream would cross its own valley, gradationally leaving gravels under sands under silts under muds under fine grains that settled in overbank floods. With nothing missing, the sequence was before us now, and was many times reiterated in the rock—a history of the migrations of the stream as it spread layer upon layer through its subsiding valley, 404, 403, 402 million years ago.

In 1832, Asher B. Durand came upon the scene. Durand was one of the founders of what in time would be labelled the Hudson River School. The term was a pejorative laid by a critic on painters who went outdoors to vent their romantic spirits. They went up the Hudson, they went up the Rockies, and they went into the Water Gap unafraid. Durand painted another Durham boat. His trees looked Japanese. The picture was published after Durand himself made a copper engraving. It contributed to the axiom that where an easel had stood a hotel would follow. Kittatinny House was established in 1833, sleeping twenty-five.

Anita chipped out a piece of Bloomsburg conglomerate—evidence in itself that the stream which had made it was by no means spent. The rolling Silurian countryside must have been lovely—its river valleys velvet green. There were highland jaspers among the pebbles in the sand.

The early geologists began arriving in 1836, led by Henry Darwin Rogers. They were conducting Pennsylvania's first geological survey. In the deep marine Martinsburg slate and in the mountain strata that stood above it—in the "plication" and the "corrugation" of the sediments—Rogers saw "stupendous crust-movement and revolution," the "most momentous" of ancient times, and reported to Harrisburg what would eventually become known as the Taconic and Alleghenian Orogenies. He decided that something had wrenched the mountain in New Jersey several hundred feet out of line with its counterpart in Pennsylvania. "I conceive these transverse dislocations to pervade all the great ridges and valleys of our Appalachian region," he wrote, "and to be a primary cause of most, if not all, of those deep notches which are known by the name of Water

Gaps, and which cleave so many of our high mountain ridges to their very bases."

There were some thin green beds among the Bloomsburg reds. Anita said they were the *Kupferschiefer* greens that had given false hope to the Dutch. Whatever else there might be in the Bloomsburg Delta, there was not a great deal of copper. In the eighteen-forties, the mines of the Minisink were started up anew. They bankrupted out in a season. The Reverend F. F. Ellinwood delivered the "Dedication Sermon" in the Church of the Mountain, village of Delaware Water Gap, Pennsylvania, August 29, 1854, a year that Ellinwood placed in the sixth millennium after Creation. "The rude blasts of six thousand winters have howled in undaunted wildness over the consecrated spot, while yet its predicted destiny was not fulfilled," he told the congregation. "But here, at length, stands, in very deed the church firmly built upon the rock, and it is our hope and prayer that the gates of Hell shall not prevail against it. . . . For many centuries past, has Jehovah dwelt in the rocky fastnesses of this mountain. Ere there was a human ear to listen, His voice was uttered here in the sighing of the breeze and the thunder of the storms, which even then were wont to writhe in the close grapple of this narrow gorge. Ere one human footstep had invaded the wildness of the place, or the hand of art had applied the drill and blast to the silent rock, God's hand was working here alone—delving out its deep, rugged pathway for

[97]

yonder river, and clothing those gigantic bluffs and terraces with undying verdure, and the far gleaming brightness of their laurel bloom." The hand of art, that very summer, was blasting the Delaware, Lackawanna & Western Railroad into the silent rock. Stagecoaches would soon leave the scene. A pathway by the river was replaced with rails. The sycamores that shaded it were felled. A telegraph wire was strung through the gap. Given a choice between utility and grandeur, people apparently wanted to have it both ways. Trains would travel in one direction carrying aristocrats and in the other carrying coal.

Anita put her fingers on fossil mud cracks, evidence not only of hours and seasons in the sun but of tranquillity in the environment in which the rock had formed. She also moved her fingers down the smooth friction streaks of slickensides (tectonic scars made by block sliding upon block, in the deforming turbulence of later times).

In the Ecological epoch, the Backpackerhaus School of photography will not so much as glance at anything within twenty-five miles of a railhead, let alone commit it to film, but in the eighteen-fifties George Inness came to the Water Gap and set up his easel in sight of the trains. The canvases would eventually hang in the Metropolitan Museum, the Tate Gallery, the National Gallery (London). Meanwhile, in 1860, Currier & Ives made a lithograph from one of them and published it far and wide. By 1866, there were two hundred and fifty beds in Kittatinny

House alone, notwithstanding that the manager had killed a huge and ferocious catamount not far from the lobby. That scarcely mattered, for this was the New World, and out in the laurel there were also wolves and bears. The gap was on its way to becoming a first-class, busy summer resort.

"Note the fining-upward cycles," Anita said. "Those are crossbedded sandstones with mud clasts at the base, rippled to unevenly bedded shaly siltstones and sandstones in the middle, and indistinctly mud-cracked bioturbated shaly siltstones with dolomite concretions at the top."

It was a lady visiting the Water Gap in the eighteen-sixties who made the once famous remark "What a most wonderful place would be the Delaware Water Gap if Niagara Falls were here."

The Aldine, in 1875, presented three wood engravings of the Water Gap featuring in the foreground gentlemen with walking sticks and ladies with parasols, their long full dresses sweeping the quartzite. The accompanying text awarded the Delaware Water Gap an aesthetic edge over most of the alpine passes of Europe. *The Aldine* subtitled itself "The Art Journal of America" but was not shy to make dashes into other fields. "The mountains of Pennsylvania are far less known and visited than many of the American ranges at much greater distance, and even less than many of the European ranges, while they may be said to vie in beauty with any others upon earth, and to have, in many sections, features of grandeur entitling them to eminent rank," the magazine told its readers. "Not only the nature lover, by the way, has his scope for observation and thought in the Water Gap. The scientist has something to do, and is almost certain to do it, if he lingers there for any considerable period. He may not have quite decided how Niagara comes to be where it is—whether it was originally in the same place, or down at the mouth of the St. Lawrence; but he will find himself joining in the scientific speculations of the past half-century, as to whether the Water Gap changed to be what it is at the Flood; or whether some immense freshet broke through the barriers once standing across the way and let out what had been the waters of an immense inland lake."

By 1877, Kittatinny House was five stories high. *Harper's Weekly*, at the end of the season, ran a wood engraving of the Water Gap in color by Granville Perkins, who had taken enough vertical license to outstretch El Greco. Under the enlofted mountain, a woman reclined on the riverbank with a pink parasol in her hand. A man in a straw boater, dark suit, was stretched beside her like a snake in the grass.

In the crossbedding and planar bedding of the Bloomsburg rocks, as we slowly traced them forward through time, there had been evidence of what geologists call the "lower upper flow regime." That was now becoming an "upper lower flow regime."

When people were bored with the river, there

were orchestras, magicians, lecturers, masquerade balls. They could read one another's blank verse:

Huge pile of Nature's majesty! how oft
The mind, in contemplation wrapt, has scann'd
Thy form serene and naked; if to tell,
That when creation from old chaos rose,
Thou wert as now thou art; or if some cause,
Some secret cause, has rent thy rocky mantle,
And hurl'd thy fragments o'er the plain below.
The pride of man may form conceptions vast,
Of all the fearful might of giant power
That rent the rampart to its very base,
Giving an exit to Lenape's stream,
And wildly mixing with woods and waters.
A mighty scene to set enchantment free,
Burst the firm barrier of eternal rock,
If by the howling of volcanic rage,
Or foaming terror of Noachian floods.
Let fancy take her strongest flight. . . .
But, as for us, let speculations go,
And be the food of geologic sons;
Who from the pebble judge the mountain's form . . .

Anita said the rock had been weakened here in this part of the mountain. The river, cutting through the formations, had found the weakness and exploited it. "Wherever a water gap or a wind gap exists, there is generally tectonic weakness in the bedrock," she went on. "The rock was very much fractured and shattered. There is particularly tight folding here."

The hotels were in Pennsylvania, and were so numerous in the eighteen-eighties and eighteen-nineties that they all but jostled one another, and suffered from the competition. Up the slope from Kittatinny House, as in a game of king-of-the-mountain, stood Water Gap House, elongate and white, with several decks of circumambient veranda under cupolas that appeared to be mansard smokestacks. All it lacked was a stern wheel. There was a fine view. On the narrow floodplain and river terraces of New Jersey, where I-80 would be, there were cultivated fields and split-rail fences, corn shocks in autumn, fresh furrows in spring.

Anita and I came to the end of the Bloomsburg, or as far as it went in the outcrops of the gap. "These are coarse basal sands," she said of one final layer. "They were deposited in channels and point bars through lateral accretion as the stream meandered." In all, there were fifteen hundred feet of the formation, reporting the disintegration of high Silurian worlds.

Ten or twelve years after the turn of the century, a Bergdoll touring car pulled into the porte cochère of Water Gap House and the chauffeur stepped out, leaving Theodore Roosevelt alone in the open back while a photograph arrested his inscrutable face, his light linen suit, his ten-gallon paunch and matching hat. This must have been a high moment for the resort community, but just as Teddy (1858-1919) was in his emeritus years, so, in a sense, was the Water Gap. A fickling clientele preferred Niagaras with falls. An intercity trolley had been added to the scene. Two miles downstream—in what had been George Inness's favorite foreground —was a new railroad bridge that looked like a Roman aqueduct. Rails penetrated the gap on both sides of the river. There was a golf course—dramatic in its glacial variations on precipitous tills pushed by the ice up the side of the broken mountain—where Walter Hagen, in 1926, won the Eastern Open Championship. Soon thereafter, the tournament was played for the last time. Walter Hagen was not coming back, and neither was the nineteenth century. The perennial Philadelphians were now in Maine. In 1931, Kittatinny House burned up like a signal fire. Freight trains wailed as they rumbled past the embers. In 1960 came the interstate—a hundred and sixty years after the first wagon road. As a unit of earth history, a hundred and sixty years could not be said to be exactly nothing—although, in the gradually accumulated red rock beside the river, ninety-four thousand such units were represented. To put it another way, in the fifteen-hundred-foot thickness of the Bloomsburg formation, there were five millimetres for each hundred and sixty years. The interstate, with its keloid configuration, was blasted into the Shawangunk quartzites, blasted into the redbeds of the Bloomsburg, along the New Jersey side. As if waves. They crest, break, and spread themselves westward. When they are spent, there is an interval of time, and then again you see the rivers running eastward. You look over the shoulder of the painter and you see all that in the landscape. You see it if first you have seen it in the rock. The composition is almost infinitely less than the sum of its parts, the flickers and glimpses of a thousand million years.

Over the bridge and out of the gap, we paid twenty-five cents at the booths of the Delaware River Joint Toll Bridge Commission. The collector was a citizen so senior he appeared to have been alive for a sixty-millionth of the history of the world. "Have a nice day," he said. We were moving west and would soon be rising into what geologists refer to as "the so-called Pocono Mountains"—actually a layered flatland that has been cut up by confused streams into forested mesas with names like Mt. Pohopoco. The long continuous welt of the deformed Appalachians—the Valley-and-Ridge belt of folded mountains—is extremely narrow at this latitude. As much as eighty miles wide in the course of its run from Alabama into Canada, it is a fifteen-mile isthmus where it is crossed by Interstate 80 at the eastern end of Pennsylvania. The foldbelt is narrow there because the Poconos refused to deform. When tectonism came and rock was being corrugated left and right, the strata that would become the Poconos were somewhat compressed but did not bend. "The rocks took the shock of the tectonics and didn't buckle," Anita said. "They shattered some, but they didn't move much. They didn't have the glide planes."

Now, scarcely a mile from the toll booths and still very much in the foldbelt, we came to what road builders call a throughcut, where the road had been blasted through the tip of a ridge. We stopped, crossed the interstate, and climbed the higher side. The rock was calcareous shale, and had been seafloor mud about three hundred and ninety million years ago, possibly ten fathoms down then, and gritty with fragments of shells and corals. There were brachiopods in the rock (something like clams and scallops) and cornucopian corals. Certain categories of these lone-growing, conical corals were the index fossils that led nineteenth-century geologists working in Devonshire to recognize the relative age of the rock the corals were found in, and to call the time Devonian. "If it weren't for this roadcut, I'd never have been able to measure these rocks," Anita said. "The next exposure is halfway between Kingston and Albany. When they first made this road, we came in and mapped in a hurry, before they laid on that goddamned grass seed, all that straw and organic tar. In the East, no one knows from geology."

The sea had been in retreat here in the early Devonian, and as we made our way uphill, and pursued the dip of the strata, we hiked two or three million years through progressively shallower marine deposits and came to a conglomerate full of pearly white quartz that had been tossed and rounded by surf. Beyond the conglomerate was light, coarse-grained sandstone—a fossil Devonian beach. The sea would have been out there to the west, the equator running more or less along the boundary between Canada and Alaska. We turned and went the other way, up through woods and around the nose of the ridge. We were far above the interstate now and looking down on the tops of big rolling boxes—North American Van Lines, moving families from coast to coast. We went on through more woods, in an easterly direction, against the dip, until at length, high on the far side of the ridge, we reached another beach, ten or fifteen million years older than the first one. In the comings and goings—transgressions, regressions—of the epicontinental sea, the strandline had paused here in late Silurian time. Anita said, "This was the barrier beach when the redbeds of the Delaware Water Gap were paper-laminated lagoon muds behind barrier islands. Geology is predictable. If you find lagoon mud, you should find beach sand not far away." On through the woods, she walked offshore to an exposure of dark, shallow-water limestone. "This is what I've come up here for," she said. "This is as pure a limestone as you can get." She cooked at high temperature. The conodonts will tell me the temperature."

In this manner, we meant to make our way west —picking and pounding at roadcuts and occasionally wandering off into the country behind them. In the deformed, sedimentary Appalachians, the rock not only had been compressed like a carpet shoved across a floor but in places had been squeezed and shoved until the folds tumbled forward into recumbent positions. Some folds had broken. Some entire regions had been picked up and thrust many miles northwest. Dozens of other complexing events had locally affected the structures of the Valley-and-Ridge terrain. One therefore could not know what to expect next. Whole sequences might suddenly be upside down, or repeat themselves, or stand on end reading backwards. Among such rocks, time moves in and out and up and down as well as by.

"It's a real schlemazel," Anita said. "Not by accident is geology called geology. It's named for Gaea, the daughter of Chaos."

Among the west-dipping Silurian formations of the Delaware Water Gap, one might project but could not reasonably expect Devonian rock to westward. It would be there if the stratigraphic package was intact and had not been overturned. The rock of that first big Pennsylvania roadcut was early Devonian in age. Leaving it, we moved seven miles west along the interstate and twenty million years up the time scale, where we stopped at a roadcut of middle

Devonian marine siltstones and shales, so rich in organic residues that it was black as carbon, with corals that had been sliced by dynamite and resembled sections of citrus. Cambrian, Ordovician, Silurian, Devonian—for this crunched and shuffled country we were experiencing remarkable consistency in an upward voyage through time. And now in the silken muds of these Devonian seafloors we were seeing the final stages in the long tranquil interval between the Taconic and Acadian revolutions. The rock coarsened abruptly as we drove on westward. There were cobbly conglomerates. They were the first explosive belch from the new Acadian mountains, which came up in the east at a rate ten times as fast as erosion could destroy them and, with a new system of rivers, rapidly shed this downpour of rock. A few miles farther on, another ten or fifteen million years, and we were among roadcuts containing upper Devonian stream channels of a quiet country, a low alluvial plain—point bars, cutbanks, ripple marks in red river sands. We were fifty million years past the Water Gap, and the geology was repeating itself on an epic scale. A new set of coalescing fans had come off the Acadian mountains, and as the great sierra disintegrated its detritus spread westward thick upon the country and into the sea—at least ten thousand feet thick in the east and gradually thinning to the west, this immense new clastic wedge, to be known in geology as the Catskill Delta.

It stands at the surface of a huge piece of country. Erosion working into the high eastern end has cut the shapes of the Catskill Mountains. The rock lies essentially flat there, and is flat all the way to the shores of Lake Erie. It is the uppermost rock of half of New York State. It is the rock of Chenango, the rock of Chautauqua, too. It is the rock of Seneca, Ithaca, Elmira, Oneonta. In Pennsylvania, it is largely buried, or was sliced and kneaded into the deformed mountains, but as the so-called Poconos it stands flat and high. The Poconos actually are part and parcel of the New York Devonian clastic wedge. The Poconos are a tongue of New York State penetrating Pennsylvania.

The Acadian mountains are gone. The wedge remains. The Acadians, in their Devonian prime, must have been a crowd of Kanchenjungas, to judge by their sedimentary remains, which reach almost to Indiana. As the mountains came down, they stood ever deeper in debris. At Denver, the Rocky Mountains are up to their hips in their own waste. The sedimentary wedge that has come off the Rockies is thickest there by the mountain front, and gradually thins to the east. Kansas and Nebraska are like pieces cut from a wheel of cheese—lying on their sides, thick ends to the west. Altitude in itself suggests the volume of material. Kansas and Nebraska are three thousand feet higher in the west than in the east.

We were running on the summits of the Poconos —uneven but essentially level topography, the Pocono equivalent of Alpine minarets. Where we saw stratified rock in roadcuts, it seemed level enough to stop a bubble. For the most part, it was Catskill sandstone, red as borscht, from latest Devonian and earliest Mississippian time. The summits of the Poconos were not only cragless, they ran on under the scrub oaks as far as the eye could see. There were peat bogs. There was a great deal of standing water. The landscape was bestrewn with hummocky lumps of gravel. "There's no way that streams brought all that gravel up here," Anita said. "Religious farmers say it's evidence of the Great Flood."

If so, the Great Flood was frozen. These were morainal gravels, outwash gravels. Interstate 80 marks almost exactly the Wisconsinan ice sheet's line of maximum advance in the Poconos.

We made a short digression from the interstate to see—in some Devonian siltstone—a tidal flat that was stuffed with razor clams. The surface of the rock had a Fulton Market look. It was a paisley of conglomerate clams. Three hundred and fifty million years old, they resembled exactly their modern counterparts. "Things haven't changed much," Anita said as she got back into the car.

We drove on into Hickory Run State Park, where we walked through heavy woods toward a clear space ahead. We seemed to be approaching a body of water. Its edge resembled a shore, and its seventeen acres were surrounded by conifers, whose jagged silhouettes invoked a northern pond. In place of water, however, the pond consisted of boulders—

thousands of big boulders, some of them thirty feet long, nearly all of them red rock weathered dusty rose, and all of them accordant with a horizontal plane, causing them to seem surreally a lake of red boulders. DAD, MOM, HARRY, and GEORGE had been there in 1970 with a can of acrylic spray. JOE VIZZARD came some years later. Dozens of others had daubed the rocks on days ranging backward to 1935, when the park was established. The big red-boulder expanse was difficult to cross, however, and its sorry guestbook was confined to one corner. The boulders were stunningly beautiful—in their lacustrine tranquillity, their lovely color, their spruce-rimmed absence of all but themselves. We walked out some distance, stepping from one to another megaton red potato.

Anita lost her balance and almost went below. "What a klutz," she remarked.

I thought I might be learning a geologic term.

"These are periglacial boulders," she went on. "They're not erratics. They haven't really moved. In the climates we have now, big boulders that are not erratics just don't appear in the woods. Only a remarkable set of conditions would produce this scene. You had to start with the right bedrock. You had to have the right angle of dip, the right erosional shape for flushing, the right distance from the glacier. The terminus of the glacier was about half a mile away. The climate was arctic. Imagine the frost heaves after water in summer got into the bedrock and that kind of winter came to explode it. Gravels, sands, and clay were completely flushed away by meltwater, leaving these boulders."

Twenty-five thousand years ago—in the late Pleistocene, or, relatively speaking, the geologic present—arctic frost had broken out the boulders and had begun the weathering that rounded them. The Acadian mountains, wearing away three hundred and fifty million years ago, had provided the material of which the boulders were made. Back on the interstate and continuing across the Pocono Plateau, we ran through more flat-lying red strata of the same approximate age, and Anita said, "Remember the Bloomsburg? This rock is fifty million years later, and it looks like the Bloomsburg, and it was formed on another low, alluviated coastal plain, when the Acadian mountains were dying down. As I've been telling you, geology is predictable once you learn a few facts. Geology repeats itself all through the rock column."

On the geologic time scale, anyone could assign these events to their respective places and sense the rhythms of the cycles—rock cycles, glacial cycles, orogenic cycles: overlapping figures in the rock. Taken all together, though, they seemed to ask somewhat more than they answered, to reveal less than nature kept concealed. The evidence showed that the Acadian mountains had come down, as had the Taconics before them, and each had spread westward new worlds of debris. One could also discern that the violence of the Acadian Orogeny had folded and faulted the sedimentary rock that had formed from the grit of the earlier mountains, and metamorphosed the rock as well—changed the shales into slates, the sandstones into quartzites, the limestones and dolomites into marble. A third revolution would follow—the Alleghenian Orogeny, in Pennsylvanian-Permian time. Another mountain wave would crest, break, and send its swash to westward. It was all very repetitive, to be sure—the great ranges rising, falling, rising, falling, covering and creating landscapes, as if successive commingling waters were to rush up a beach and freeze. But why? How? You see in rock that geology repeats itself, but you do not see what started the process. In the rivers in rock you find pieces of mountains, but you do not find out why the mountains were there.

I said to Anita, "What made the mountains rise?"

"The Acadian mountains?"

"All of them—Taconic, Acadian, Alleghenian. What made them come up in the first place?"

It has been Anita's style as a geologist to begin with an outcrop and address herself to history from there—to begin with what she can touch, and then to reason her way back through time as far as she can go. A river conglomerate, as tangible rock, unarguably presents the river. The river speaks of higher ground. The volume of sediment that the river has carried can imply a range of mountains. To find Precambrian jaspers in the beds of younger rivers means that the Precambrian, the so-called basement rock, was lifted to form the mountains. These are sensible inferences drawn cleanly through an absence of alternatives. To go back in this way, retrospectively, from scene to shifting scene, is to go down the rock column, groping toward the beginning of the world. There is firm ground some of the way. Eventually, there comes a point where inference will shade into conjecture. In recesses even more remote, conjecture may usurp the original franchise of God.

By reputation, Anita is a scientist with an exceptionally practical mind, a geologist with few weaknesses, who is at home in igneous and metamorphic petrology no less than in sedimentology. She has been described as an outstanding biostratigrapher, a paleontologist who knows the rocks in the field and can go up to a problem and solve it. In my question to her, I was, I will confess, rousing her a little. I knew what her answer would be. "I don't know what made the mountains come up in the first place," she said. "I have some ideas, but I don't know. The plate-tectonics boys think they know."

When the theory of plate tectonics congealed, in the nineteen-sixties, it had been brought to light and was strongly supported by worldwide seismic data. With the coming of nuclear bombs and limitation treaties and arsenals established by a cast of inimical peoples, importance had been given to monitoring the earth for the tremors of testing. Seismographs in large numbers were salted through the world, and over a decade or so they revealed a great deal more than the range of a few explosions. A global map of earthquakes could be drawn as never before. It showed that earthquakes tend to concentrate in lines that run up the middles of oceans, through some continents, along the edges of other continents—seamlike, around the world. These patterns were seen—in the light of other data—to be the outlines of lithospheric plates: the broken shell of the earth, the twenty-odd pieces of crust-and-mantle averaging sixty miles thick and varying greatly in length and breadth. Apparently, they were moving, moving every which way at differing speeds, awkwardly disconcerting one another—pushing up alps—where they bumped. Coming apart, they very evidently had opened the Atlantic Ocean, about a hundred and eighty million years ago. Where two plates have been moving apart during the past five million years, they have made the Red Sea. Ocean crustal plates seemed to dive into deep ocean trenches and keep on going hundreds of miles down, to melt, with the result that magma would come to the surface as island arcs: Lesser Antilles, Aleutians, New Zealand, Japan. If ocean crust were to dive into a trench beside a continent, it could lift the edge of the continent and stitch it with volcanoes, could make the Andes and its Aconcaguas, the Cascade Range and Mt. Rainier, Mt. Hood, Mt. St. Helens. It was a worldwide theory—revolutionary, undeniably exciting. It brought disparate phenomena into a single story. It explained cohesively the physiognomy of the earth. It linked the seafloor to Fujiyama, Morocco to Maine. It cleared the mystery from long-known facts: the glacial striations in rock of the Sahara, the equator's appearances in Fairbanks and Nome. It was a theory that not only opened oceans but closed them, too. If it tore land apart, it could also suture it, in collisions that perforce built mountains. Italy had hit Europe and made the Alps. Australia had hit New Guinea and made the Pegunungan Maoke. Two continents met to make the Urals. India, at unusual speed, hit Tibet. Eras before that, South America, Africa, and Europe had, as one, hit North America and made the Appalachians. The suture was probably the Brevard Zone, a long, northeast-trending fault zone in the southern Appalachians with very different rock types on either side and no discernible matchup of offset strata. Discontinuous extensions of the Brevard Zone seemed to reach to the Catoctin Mountains and on to Staten Island. Southeastern Staten Island apparently was a piece of the Old World. Ships that sailed for Europe had arrived when they went under the bridge.

Plate theory was constructed in ten years by people with hard data who were consciously and frankly waxing "geopoetical" as well. Once the essentials of the theory were complete—after the discovery of seafloor spreading had led to an understanding of trench subduction, and after the plates and their motions had finally been outlined and described—the theory took a metaphysical leap into the sancta of the gods, flaunting its bravado in the face of Yahweh. It could make a scientist uncomfortable. Instead of reaching back in time from rock to river to mountains that must have been there—and then on to inference and cautious conjecture in the dark of imperceivable unknowns—this theory by its conception, its nature, and its definition was applying for the job of Prime Mover. The name on the door changed. There was no alternative. The theory was panterrestrial, panoceanic. It was the past and present and future of the world, sixty miles deep. It was every scene that ever was on earth. Either it worked or it didn't. Hoist it was to Heaven with its own petard. "Established" there, it looked not so much backward from the known toward the unknown as forward from the invisible to its product the surface of the earth. Anita was more worried than made hostile by all this. By no means did she reject plate theory out of hand. There were applications of it with which she could not agree. Moreover, it was too fast a vehicle for its keys to be given to children.

·   ·   ·

"The plate-tectonics people have certain set patterns that they expect to see," Anita said. "They kind of lock themselves in. If something doesn't fit the theory, they'll find some sort of reason. They'll say that something is missing, or that it was subducted, or that it has not yet been found in the subsurface. They make things fit."

"Do you believe that ocean crust is subducted into trenches, that it melts and then comes up behind the trenches as volcanoes and island arcs?" I asked her.

"That is straightforward," she said. "And I have no doubt that one edge of the Pacific Plate is grinding northwest through California. What I object to is plate tectonics taken as absolute gospel. To stuff that I know about, it's been overapplied—without attention to geologic details. It's been misused terribly. It has misrepresented facts. It has oversimplified the world. The Atlantic spreading open I absolutely believe. How long it has been spreading open I don't know. I don't really believe that North America and South America were up against one another. The whole Pacific margin is thrusting from west to east, but there is no continent colliding with it. I don't see that plate tectonics explains all of these things. I think tectonics on continents is different from tectonics in oceans, and what works in oceans is often misapplied on land. As a result, there is less understanding of regional geology. The plate-tectonic model is so generalized and is used so widely that people do not get good regional pictures anymore. People come out of universities with Ph.D.s in plate tectonics and they couldn't identify a sulphide deposit if they fell over it. Plate tectonics is not a practical science. It's a lot of fun and games, but it's not how you find oil. It's a cop-out. It's what you do when you don't want to think."

Before the plate-tectonics revolution, back in the penumbra of what is called the Old Geology, mountains were thought to be driven upward from a deep-seated source known as a geosyncline. It was a profound downwarp of the crust, a long trough below the sea, and sediments fell into it, in U-shaped configuration. East of North America, for example, the muds that would become Martinsburg slates first became rock in a geosyncline. The great trough trended northeast, like the mountains it would produce. How the mountains came up was not absolutely defined. Isostasy was presumed to participate: place enough weight on the earth's mantle and the mantle would kick like a trampoline. In addition, "earth forces" came up under the geosyncline and assisted its ascent to the sky. The earth forces were "not well understood." Like the drying skin of an apple, the uplifted material wrinkled into mountains. The story seemed clear, even if the authorship was somewhat moot, and it was a story of rhythmically successive orogenies, chapter headings in the biography of the earth. Some geologists preferred to liken them to punctuation marks, because mountain-building phases took up so little of all time—as little as one per cent, no more than ten per cent, depending on the geologist who was calculating the time.

Anita said, by way of example, "The Gulf of Mexico is a big geosyncline, if you want. The big bird-foot delta of the Mississippi River is one hell of a sedimentary pile. Drill twenty-two thousand feet down and you're still in the Eocene. The crust will take about forty thousand feet of sediment—that's the elastic limit. Then it regurgitates the sediment, which begins to rebound. The sediment is also heated up, melted. Water, gas, and oil come out of the rock. Sedimentary layers move up with thermal drives as well as with isostasy. Sedimentary layers also move laterally, and are thus thrust sheets. In Cambro-Ordovician time, fifty million years before the Taconic mountains came up, the continent was to the west of us here, the coastline was in central Ohio, and to the east of us, where the Atlantic shelf is now, stood an island arc like Japan. There are volcaniclastic sediments of that age from Newfoundland to Georgia—just about the length of Japan. The present coast of Asia is the Ohio coastline in that story. Picture the sediment that is pouring off the Japanese islands into the Sea of Japan. The Martinsburg slates were shed not from the continent—not from Ohio—but mainly from the east, from the island arc offshore. You pile up forty thousand feet of sediment and it pops. The Martinsburg popped. The Taconic mountains came up. Once the process starts, it keeps itself going. You push up a mountain range, erode it into the west. The material depresses the crust. It is low-density material and it is brought down into the regime of high-density material. When enough has been piled on, the low-density material comes back up. That is how orogenic waves propagate themselves, each mountain mass being cannibalized to produce a new mountain mass to the west. But I still don't know what started the process."

Historical geologists, in the olden days, pieced together that narrative. Economic geologists, in their pragmatic way, cared less. In describing the minable Martinsburg—the blue-gray true unfading slate— C. H. Behre, Jr., wrote in 1933, "Sedimentary rocks are often compressed from the sides through what may be loosely described as shrinking of the crust of the earth; how this shrinking is brought about is, for the present purpose, beside the point. It has the well-recognized effect, however, that layers or bedding planes are wrinkled or thrown into 'folds.'"

By the nineteen-sixties, what Behre had loosely described was widely believed to be the impact of one continent colliding with another, as Iapetus, the proto-Atlantic ocean, was closed and the suture of the two continents became the spine of the Appalachians. The successive pulses of orogeny— Taconic, Acadian, Alleghenian—were attributed to the irregular shapes of shelves and coastlines of the continents. Where they bulged, the action would have an early date, and especially where some cape, point, or peninsula had a similar feature coming from the opposite side. Such headlands, in advance contact, were said to have produced the Taconic

Orogeny. Great bays, eventually coming against one another, set off the Acadian Orogeny. The Alleghenian Orogeny was the final crunching scrum, completing the collision. The apparent suture was a line running through Brevard, North Carolina, more or less connecting Atlanta, Asheville, and Roanoke, not to mention Africa and America.

The Martinsburg seafloor and the underlying carbonate rocks had unquestionably been broken into thrust sheets and shuffled like cards. Uplifted with their Precambrian basement, they had, in perfect harmony with the Old Geology, become mountains that shed their sediments—shed their clastic wedges—and buried the Martinsburg deep enough to turn it into slate, buried the carbonates deep enough to turn them into marble. Thus, plate tectonics fit. Plate tectonics may have restyled the orogeny and dilapidated the geosyncline, but it fit the classical evidence.

There were, to be sure, certain anomalies, which suggested further study. If the Brevard Zone was the suture, how come it was so short? It was evident for a hundred miles, dubious for a few hundred more, and nonexistent after that. If the Taconic, Acadian, and Alleghenian Orogenies were subdivisional impacts of a single intercontinental collision, how come they took so long? In plate-tectonic theory, plates move at differing speeds, the average being two inches a year. The successive orogenic pulses that resulted in what we know as the Appalachian Mountains occurred across a period of about two hundred and fifty million years. In two hundred and fifty million years, at two inches a year, you can move landmasses a third of the way around the world. Geologists ordinarily require vast stretches of time to account for their theories. In this case, they have too much time. They have two continents in the act of collision for two hundred and fifty million years.

In 1972, scarcely four years after the lithospheric plates had first been identified and the theory that described them had become news in the world, Anita and two co-authors published a set of papers offering signal evidence of plate tectonics in action, apparent proof that Sweden, or something like it, had once been in Pennsylvania. It was an inference drawn from conodont paleontology. The papers were widely cited for their support of plate-tectonic theory, and are cited still. North of Reading, in the Great Valley of the Appalachians, Anita had found early Ordovician conodonts of a type previously unknown in North America but virtually identical with early Ordovician conodonts from the rock of Scandinavia. All over North America were early Ordovician conodonts from tropical and subtropical seas. Their counterparts in Scandinavia were from cooler water, and so were these strangers Anita found in Pennsylvania. She found them in what is known as an exotic block, embedded in a far-travelled thrust sheet. They happened to be within a third of a mile of warm-water American conodonts in rock of about the same age which had moved hardly at all. The Scandinavian conodonts had apparently come to Pennsylvania with the closing of the proto-Atlantic ocean and been dropped ashore off the leading edge of the arriving plate. "Even I said, 'Oh, this piece peeled off the oncoming European-African plate and got dumped in here along with the clastics,'" Anita said, telling the story. "Everybody cited those papers. To this day they are called 'marvellous, landmark papers.' I could eat my heart out. The papers have been used as prime backup proofs of plate-tectonic applications in the northern Appalachians. Even now, a lot of the people who use the plate-tectonic model for interpreting the Appalachians are completely unaware that those papers were based on a paleontological misinterpretation."

Working in Nevada three years later, Anita had found Scandinavian-style conodonts of middle Ordovician age. Her husband, Leonard Harris, savored the discovery not for its embarrassment to his wife, needless to say, but for its air-brake effect on the theory of plate tectonics. "Now, how could that be?" he would ask. "How did this happen? Europe can't hit you in Nevada." From the Toiyabe Range she had taken the cool-water fossils, and moving east to other mountains—from basin to range—she had come to middle Ordovician carbonates that contained a mixture of conodonts of both the cool-water and the warm-water varieties, the American and the Scandinavian styles. Farther east, in limestone of the same age in Utah, she had found only warm-water conodonts. She realized now the absoluteness of her error. Utah had been pretty much the western extremity of the vast Bahama-like carbonate platform that covered North America under shallow Ordovician seas. In western Utah, the continental shelf had begun to angle down toward the floor of the Pacific, and in central Nevada the continent had ended in deep cool water. The conodont types differed as a result of water temperatures, not as a result of their geographic origins. Shallow or deep, conodonts of northern Europe were the same, because the water was cool at all depths. But here in America, with the equator running through the ocean where San Francisco would someday be, Ordovician water temperatures varied according to depth. Those apparently Scandinavian fossils were forming in deep cool water, the American ones in warm shallows.

Moving east from the Toiyabe Range and into Utah, Anita had gone from outcrop to outcrop through the Ordovician world, from ocean deeps to the rising shelf into waist-deep limestone seas. She could see now that the thrusting involved in the eastern orogenies had shoved the cool-water conodonts and their matrix rock from the deep edge of the continental rise into what would be Pennsylvania. They had travelled, to be sure—but they had more likely come from Asbury Park than from Stockholm. In the thrusting and telescoping of the strata, the transition rocks of the proto-Atlantic's eastern slope had been deeply buried. In them, almost surely, would be a mixture of cool-water and warm-water conodont types. To the east of the Toiyabe Range, there had been less telescoping, and the full sequence was traceable—from the cool deep continental edge up the slope to the warm far-reaching platform. "The change had nothing to do with moving plates," Anita concluded. "Nothing to do with plate tectonics. I blew it. It was an environmental change, an environmental sequence."

More recently, working in Alaska, she had seen the sequence again, this time in tightly banded concentration, for the "American" conodonts were from reefs around Ordovician volcanic islands with steeply plunging sides and the "Scandinavian" conodonts were from cold deeps nearby.

Swerving to avoid a pothole, Anita said, "The plate-tectonics boys look at faunal lists and they go hysterical moving continents around. It's not the paleontologists doing it. It's mostly the geologists, misusing the paleontology. Think what geologists would make of the present East Coast of the United States if they did not understand oceanography and the resulting distribution of modern biota. Put yourself forty or fifty million years from now trying to reconstruct the East Coast of the United States by looking at the remains in the rock. God help you, you would probably have Maine connected to Labrador, and Cape Hatteras to southern Florida. You'd have a piece of Great Britain there, too, because you see the same fauna. Well, did you ever hear of ocean currents? Did you ever hear of the Gulf Stream? The Labrador Current? The Gulf Stream brings fauna north. The Labrador Current brings fauna south. I think that a lot of the faunal anomalies you see in the ancient record, and which are explained by invoking plate tectonics, can be explained by ocean currents bringing fauna into places they shouldn't be. In the early days of plate tectonics, a lot of us, including me, jumped on the bandwagon in order to explain the distribution anomalies we were seeing not only in the eastern Appalachians but in North America as a whole. When we better understood the paleoecologic controls on the animals some of us were working on, there was no reason to invoke plate tectonics."

The experience was cautionary, to say the least. It did not close her mind to plate tectonics, but it opened a line of suspicion and made her skeptical of the theory's insistent universality. Her discomfort varies with distance from the mobile ocean floors. She likes to describe herself as a "protester." The protest is not so much against the theory itself as against excesses of its application—up on the dry land. "A number of these people took very interesting ideas that apply to ocean floors and tried to apply them to everything," she remarked. "They tried to extrapolate plate tectonics through all geologic time. I don't know that that holds. My husband has blown some of their ideas apart."

Leonard Harris, sometimes known as Appalachian Harris, was very much a protester too. Tragically, he died in 1982, a relatively early victim of cancer and related trouble. He was a genial and soft-spoken, almost laconic man with a lean figure that had walked long distances without the help of trails. He liked to build ideas on studied rock, and was not easily charmed by megapictures global in their sweep. He referred to the long deep time before the Appalachian orogenies as "the good old days." With regard to plate tectonics, he looked upon himself as a missionary of contrary opinion—not flat and rigid but selective, where he had knowledge to contribute. His wife has compared him to Martin Luther, nailing theses to the door of the castle church.

For some years he assisted oil companies in the training of geologists and geophysicists in southern-Appalachian geology, and in return the companies made available to him their proprietary data from seismic investigations of the Appalachian crust. Over the past couple of years, these data have been supplemented by the seismic thumpings of the U.S.G.S. and COCORP—the Consortium for Continental Reflection Profiling—whose big trucks go out with devices that literally shake the earth while vibration sensors record wave patterns reflected off the rock deep below. The technique is like computed axial tomography—the medical CAT scan. The patterns reveal structure. They reveal folds, faults, laminations, magmatic bodies both active and cooled. They report the top of the mantle. They also reveal density, and hence the types of rock. Moving cross-country, the machines make subterranean profiles known as seismic lines. Seismic shots from exploding dynamite have been used for years in the search for oil. Alaska is crisscrossed with all but indelible "seismic" disruptions of tundra. The reserves of Prudhoe Bay were discovered in this manner. Dynamite in the populous East could irritate the public, so COCORP and the U.S.G.S. have adopted a behemoth called Vibroseis to shiver the timbers of the earth. One of the first discoveries the vibrations reported was that the Brevard Zone is relatively shallow and the crust below it is American rock that does not in any vague way reflect a continent-to-continent suture. Africa was nowhere in the picture. The Brevard Zone proved to be the tobogganlike front end of a large and essentially horizontal thrust sheet.

Plate-tectonic theorists accommodated this news by moving the suture fifty miles east. The new edge of Africa was under Kings Mountain. Seismic shots took the stitches out of Kings Mountain. "When we got the data for the Brevard"—as Leonard Harris liked to tell the story—"they pushed the suture to Kings Mountain, and when we got data for that they said the suture must be under the coastal plain, and now that we are getting data for the coastal plain they say it must be in the continental shelf. Well, we've got data out there, too." Up and down the Appalachians, wherever such data were collected thrust sheets were seen to have moved in a northwesterly direction, and much of the thrusting had never been suspected before. Conventional thought had been that the old rock of the Green Mountains, the Berkshires, the New Jersey Highlands, the Catoctin Mountains, and the northern Blue Ridge was in place, firmly rooted—autochthonous, as geologists are wont to say. It may have been crushed and pounded in the various orogenies, and metamorphosed, too, but it was nonetheless thought to be securely glued where it first had formed as rock. The belt was supposed to have been the fixed starting block from which, somehow, thrusting had proceeded northwest. The idea had come up through the Old Geology and been incorporated into the substance of plate tectonics. Then, in 1979, Vibroseis rumbled into the country and showed that from Quebec to the Blue Ridge the entire belt was deracinated. The Great Smokies and the Skyline Drive, Camp David and the Reading Prong, the Berkshires and the Green Mountains—all of it had moved, at least a few tens of miles and as much as a hundred and seventy-five miles, northwest.

Using the new data, Leonard meticulously drew a palinspastic reconstruction of North American rock, showing it as it had appeared before it was shoved and deformed. He chose a cross-section that had been shot more or less from Knoxville to Charleston and out to sea. The reconstruction showed that the rock of the Valley and Ridge—the folded-and-faulted, deformed Appalachians—had been squeezed so much that its breadth had been reduced about sixty miles. The supposedly rooted Blue Ridge had been moved inland from what is now the coast. Rock of the present Piedmont had come from three hundred miles out in the present sea. This left Africa out in the cold and plate-tectonic theory in no small need of a substitute for what had been— and in many classrooms would continue to be—the world's most "classical" example of a continent-to-continent suture.

With patience geological, the believers restyled their belief, apparently according to the criterion "If at first it doesn't fit, fit, fit again." There was suggestive help from the West. A great deal of land out there had not been there when the carbonate rocks sloped away to ocean-crustal deeps in Ordovician time. California, Oregon, Washington, British Columbia had appeared where there was no continental structure of any kind. Up and down the western margin, in fact, there was an unaccounted-for swath of land averaging four hundred miles wide. There was also the whole of Alaska. How did all that country come to be where it is? What compressed the western mountains? If Europe were on the international date line, these questions would have a ready answer, but inconveniently it was not. No one was enthusiastic enough to suggest a hit-and-run visit from China. Where, then, since Ordovician time, had the North American continent acquired nine hundred million acres of land?

There was an answer in the concept of microplates, also known as exotic terrains. New Guineas, New Zealands, New Caledonias, Madagascars, Kodiaks, Mindanaos, Fijis, Solomons, and Taiwans had come over the sea to collect like driftwood against the North American craton. The first such terrain identified was Wrangellia, named for the stratovolcanic Wrangells, some of the Fujis of Alaska. Dozens of other exotic terrains have since been named—Sonomia, Stikinia, the Smartville Block. If a piece of country is possibly exotic and possibly not—if it is so enigmatic that no one can say whether it has come from near or far—it is known as suspect terrain. I came home not long ago from a visit to the country north of the Tanana River, in eastern interior Alaska, where streams that resemble gin come down from mountains and into the glacial Yukon. A geologist in New Jersey welcomed me home with an article from the British journal *Nature* which described the Alaskan region of the upper Yukon. "The terrane is probably composite," said *Nature*, "with nappes of upper Palaeozoic oceanic assemblages thrust across a quartzo-feldspathic and silicic volcanic-rich protolith of probable Precambrian to known Palaeozoic age and of unknown continental affinity." I was appalled to discover that that was where I had been, and mildly disturbed to learn that terrain long familiar to me had now become suspect.

(Geologists write "terrain" when they mean topography and "terrane" when they are referring to a piece of country many miles deep. I am not a geologist and I refuse to coöperate.)

Taiwan, at this writing, is evidently on its way to the Chinese mainland. Taiwan is the vanguard of a lithospheric microplate and consists of pieces of island arc preceded by an accretionary wedge of materials coming off the China Plate and materials shedding forward from the island's rising mountains. As the plate edges buckle before it, the island has plowed up so much stuff that it has filled in all the space between the accretionary wedge and the volcanic arc, and thus its components make an integral island. It is in motion northwest. For the mainland government in Peking to be wooing the Taiwanese to join the People's Republic of China is the ultimate inscrutable irony. Not only will Taiwan inexorably become one with Red China. It will hit into China like a fist in a belly. It will knock up big mountains from Hong Kong to Shanghai. It is only a question of time.

As an exotic terrain on the verge of collision with a continent, Taiwan is a model not only for the building of the American West but for the application of microplate-tectonic theory to the eastern orogenies and the closing of the proto-Atlantic. In this respect, a plane fare to Taiwan has been described as "a ticket to the Ordovician," a time when something or other, beyond question, produced the

Taconic Orogeny, and if it was not the slamming-in of a continent against North America, then possibly it was the arrival of an exotic island like Taiwan. The analogy becomes wider. South of Taiwan are Luzon, Mindanao, Borneo, Celebes, New Guinea, Java, and hundreds of dozens of smaller islands from the Malay Peninsula to the Bismarck Archipelago. Coming up below them is Australia, palpably moving north, headed for collision with China, with a confusion of microplates lying between. According to microplate theory, as Europe, Africa, and South America closed in upon North America through Paleozoic time, there rode before them an ocean full of Javas, New Guineas, Borneos, Luzons, Taiwans, and maybe hundreds of dozens of smaller islands. The Avalon Peninsula of Newfoundland appears to have been a part of such an island, and the Carolina Slate Belt, and a piece of Rhode Island east of Providence, and Greater Boston. A schedule of arrivals of incoming exotic terrains will account—as a simple continent-to-continent collision cannot do—for the long spreads of time between one and another of the Appalachian mountain-building pulses. As someone once compacted it for me, "you sweep the New Zealands and Madagascars out of the ocean and then you close it with the Alleghenian Orogeny." Disagreeing interpreters see terrains of highly varied dimension. Nominated as the terrestrial remains of one exotic block is the whole of New England from Williamstown eastward, arriving in the Ordovician to lift the Taconic mountains.

Exotic terrains and their effects represent only one of the responses of plate-tectonic theorists to the embarrassment caused by the failure of Exhibit A among intercontinental collisions to exhibit a suture. Another response has been the notion that when two continents collide there is every possibility that one will split the other, like an axe blade entering cedar; if so, you would find the invaded country rock both above and below the invader. The concept is known as flake tectonics. Its message to Vibroseis is to stop shaking and go home. With a little erosion and flake tectonics, you can have the native rock reaching far under the rock from across the sea. Even so, the bunching of exotic terrains seems to solve more problems than flake tectonics does. Exotic terrains not only explain the intervals of time involved in the Taconic, Acadian, Alleghenian Orogenies, they suggest as well why Taconic deformation occurred in the northern but not the southern Appalachians. Shortening collisional boundaries, they restore some dignity to the Brevard suture.

Anita turned on the windshield wipers and wiped an April shower. Beside the interstate, the Pocono Devonian roadcuts were of much the same age and character as ones we had seen before. We passed them by. "Better not to do geology in the rain," Anita said. "It's unfair to the rocks." With regard to the possibility of exotic terrains having added themselves to eastern North America, she said, "If you stretch out the overthrusts in the Appalachians, they show that—before the mountain-building began—the continent was much larger than it is now, not smaller."

I remembered Leonard Harris—one day at their home in Laurel, Maryland—saying, "The Brevard Zone is the sort of fault you would see in any thrust belt. With the plate-tectonic model, anybody can write a history of an area without having been there. These people have no way to evaluate what they're doing. They just make up stories."

"Plate-tectonic interpretations often start where data stop," Anita had said. "These people will just *float* microplates around. If the West is made of microplates, where the hell was the landmass that produced the pieces?"

"They want to be science-fiction writers," Leonard said. "That's what they want to do. They really look at it in a science-fiction mode. I have never been able to do that. If you don't know what caused something, you don't know; and that's the way it is."

"Yeah, but it's a much more romantic way to look at things," she said. "And it certainly does turn students on."

"People love it."

"It allows them to play all kinds of games without the necessity and painstaking dogwork of gathering facts. It allows them to write papers without killing themselves getting data."

"People want the science-fiction story. It's easier to believe that pieces of the world move than it is to see a sand grain move. The principal problem about interpreting the Appalachians is that there have been no available subsurface data in the Blue Ridge and Piedmont. All interpretations, up until 1979, were based on what people thought was a rooted system. Their ideas were based on offshore data, where they had 3-D—you know, seismic data, magnetic data— and these data were more or less applied onshore. The concepts were developed from the ocean to the land. Now that we are beginning to get subsurface data on land, we are testing their concepts. A lot of what people have been saying is not hanging together. Some of what they have said *has* hung together."

I said to them, "One would gather from the seismic lines that for a continent-to-continent collision you'd have to go pretty far east to find the suture."

"I don't think you can go far east enough," Anita said. "The oceanic basin is out there."

"When you start working on the shore and you look offshore, you've got an immediate problem," Leonard said. "They tell us that the oldest ocean crust that has ever been found is Jurassic—a hundred and fifty million years old. Onshore, we have everything that's ever been built—from the Precambrian on up. We have a continuum. We have something that has been preserved much longer than a hundred and fifty million years. We have rock that is three-point-eight billion years old. So we have a problem relating. If all the ocean crust is Jurassic, or younger, there's a lot happening here onshore that is never preserved out there. It's difficult to compare the two."

Anita said, "I believe in plate tectonics—just not in the way they're perpetrating it for places like the East Coast. It shouldn't be used as the immediate answer to every problem. That's what I object to. Now that their suture zones have disappeared, people are going to microplates."

"They seem to be saying that you don't have to see any order," Leonard said. "Because it's all chaos, and if it's chaos why worry about it?"

"What we try to do is pull the thrust plates apart and make them into some sort of recognizable geologic model," Anita said.

Leonard said, "You pull something apart to see what it might have been, not what you think it was in advance. It might have been a shelf, a basin. You work at it, and see what it was."

"The plate-tectonics boys make no attempt to do this, because they see no reason to," Anita said. "There are too many pieces missing. Each existing piece is an entity unto itself. Everything is random pieces."

"Most people have never had an opportunity to work with thrust-faulted areas. We've lived with them all our lives. If we go along a fault system far enough, we can actually see the next thrust plate. Maybe I'll have to go a hundred miles until I find out what it really looked like. You do that by making a model. You pull the thrusts apart and see what the country originally looked like. But until you've done that, and been faced with that problem, it's natural to say, 'God, these are so different. They could be microcontinents.' You can reconstruct a large flat piece out to the east as an original depositional basin. You can see volcanic terrain that was partly onshore, partly offshore. You can look at that as a basin, too, just sitting there, a continuous thing. You see the same thing from Georgia north. The Appalachian belts are almost continuous basins, showing different kinds of depositional patterns. They're not exotic pieces."

"Not at all."

"Science is not a detached, impersonal thing. People will be influenced as much by someone who is a spellbinder as by someone with a good, logical story. It is spellbinding to say that these belts are exotic and were built through time by micro or macro pieces aggregated to the continent. But the fact that you've got seismic lines without any apparent suture lines makes you wonder what really happened. Where are those Devonian and Taconic sutures? Are they just not being recognized? Or are they in fact thrust plates?"

.　　.　　.

I thought also of field trips in the company of geologists trying to puzzle out the details of plate-tectonic theory. Metamorphic details. Geophysical details. The dialogue is not without crescendos. They debate in a language exotic in itself, and shuffle like a blackjack deck the stratigraphic units of the world. In Vermont, say, walking the hard-packed dirt roads among Black Angus meadows and roll-mop hay, over plank bridges—"LEGAL LOAD LIMIT 24,000 POUNDS"— and down through the black spruce to Cambrian outcrops jutting up as ledges in fast, clear streams, they argue.

"You've got the right first approximation, but you've got to go ahead and prove that it's the correct approximation."

"We're talking about developing fabrics."

"It's pretty clear now that fabrics don't develop that way."

"For an anisotropic crystal, I don't think you can say what you just said. You've got to put in another sentence there to justify using that approximation."

"I don't see what you're saying. Anisotropic or not, that's the definition of being stressed."

"When you say the thermodynamic stability of the phase that's growing is sigma 1, sigma 2, sigma 3, divided by 3, it's proof for an isotropic crystal. For garnets growing, that's fine. For mica, that's not fine."

The hills roundabout are decidedly footloose and no one knows how far they have moved. The rock they are made of has flopped over in recumbent folds and is older than the rock it rests on. In the Old Geology, these hills were described as large pieces of

the high Taconic mountains, which had slid down-
hill by gravity and come to rest in the westward seas.
Now they are seen variously as remnants of thrust
sheets or as a possible exotic terrain.

"You can perhaps picture for yourself that the
allochthon was at one time more extensive. It was
coming this way through a sea of black mud, and
here is the record of it. This is where it touches
North America—at least that's a possibility."

"But there are no remnants of the western side
of the ocean."

"Evidence would be in the rest of the conglomerate, little bits of limestone debris. Evidence
would be in the seismic line."

"But that evidence could be . . . You could im-
bricate the stuff that's coming off North America."

"Yes. Yes, you can."

"So I don't think that's definitive."

"I'm not saying it is. I'm just saying here's another possibility. And I'm going to stick to that for
the time being, as well as the Chain Lakes ophiolite."

It doesn't matter that you don't understand
them. Even they are not sure if they are making
sense. Their purpose is trying to. Everyone has
crowded in. The science selects these people—with
their jeans and boots and scuffed leather field cases
and hats of railroad engineers. To them, just being
out here is in no small measure what it's all about.
"The three key things in this science are travel,
travel, and travel," one says. "Geology is legitimized

tourism." When geologists convene at an outcrop, they see their own specialties first, and sometimes last, in the rock. People listen closely for techniques applicable to areas they work in elsewhere. If someone is a specialist in little bubbles that affect cleavage planes, others will turn to the specialist for comment when cleavage is of interest in the rock. The conversation runs in links from specialty to specialty, from minutia to minutia—attempting to establish new agreement, to identify problems not under current research. From time to time, details compose. The picture vastly widens.

"Aren't we in North America?"

"You are in North America. Yes."

"And you are in Europe."

"That's one possibility. Yes."

"You are standing across the ocean."

"No. I'm not standing across the ocean. I'm transplanted here. Is the Atlantic between me and you? No."

"You are allochthonous."

"You're damned right I am."

"You are rootless."

"Not to mention recumbent."

"Only after hours."

"There may be another suture."

"There may be another suture, but this is the only one we've got."

"No, no, you've got another one, which goes up through Quebec."

"No, that's not in place. The Canadian seismic line proves it. You will remember also that Laval—way back, 1965—came out with late Ordovician fossils in the Sherbrooke anomalies. Where else do you find a continuous sedimentation from the middle Ordovician up to the Silurian—from Rangeley Lake up to New Brunswick in one belt? The Sherbrooke thing, restored, would come from where it ought to come from. So I'm suggesting either that two continents collided, and you have one basin there, receiving continuous sedimentation, or . . ."

"You may have a double arc."

"You might have a double arc."

"There's another solution."

"Sure, but I'm saying let's take the simplest configuration."

"Why not just have one arc with basins on both sides of it?"

"No. No. You have the Bronson Hill anticlinorium, and then you have the Ascot Reef."

"You have a volcanic arc on the stable side of the subduction zone, an expected arc above the downgoing slab."

"You have a short-lived slab going down below the Ascot Reef and the one of longer duration that's on the other side. I would somehow think that there has to be something in these rocks, in the limestones, that you'd be able to hopefully connect to that platform."

"The only thing I can say is . . ."

"What about the blue quartz?"

"What about the blue quartz? The stratigraphy of the Taconic rock matches unit for unit with Cambrian rocks of Avalon, and the fossils look alike. That's all I can tell you. Nowhere else do we find this sort of thing except Wales."

A structural geologist with a foot on each continent looks up and aside from this contentious scene. "While geologists argue, the rocks just sit there," he remarks. "And sometimes they seem to smile."

.   .   .

The car hit an erosional vacuity that almost threw it off the road. Geology versus the State of Pennsylvania. Geology wins. In Eastern weather, the life expectancy of an interstate is twenty years. Mile after mile, I-80 had been heaved, split, dissolved, and cratered. A fair amount of limestone is incorporated in the road surface in Pennsylvania. Limestone is soluble in distilled water, let alone in acid rain. "Acid rains eat the surface, then water goes in and freezes, thaws, freezes again, and fractures the hell out of the road," Anita said, easing down toward minimum speed. "That, of course, is exactly how water works on bedrock. But an interstate can't be compared to bedrock. An interstate has no soil protecting it. And it's mostly carbonate. It's not very resistant stuff."

We were sixty miles into Pennsylvania and had

descended from the Pocono Plateau, generally running backward through time and down through the detritus of two great ranges of mountains. Now the country was familiar—valley, ridge, valley, ridge. We were again in the deformed Appalachians. While the Delaware Water Gap had been a part of the main trunk of the foldbelt, this was an offshoot that curled around the western Poconos—a broad cul-de-sac whose long ropy ridges ended like fingers, gesturing in the direction of New York State. It was rhythmic terrain, predictable and beautiful, the quartzite ridges and carbonate valleys of the folded-and-faulted mountains, trending southwest, while the interstate negotiated with them for its passage toward Chicago. Looked at in continental scale on a physiographic map or a geologic map—on almost any map that doesn't obscure the country with exaggerated human improvements—the sinuosity of the deformed Appalachians is as consistent as the bendings of a moving serpent. In Alabama, the mountains come up from under the Gulf Coastal Plain and bend right into Georgia and then left into Tennessee and right into North Carolina and left in the Virginias and right in Pennsylvania and left in New Jersey and New York and right in Quebec and New Brunswick and left in Newfoundland. Some people believe that in this Appalachian sinuosity we are seeing the coastline of the Precambrian continent—North America in the good old days, when the Taconic Orogeny was off in the future and these big, scalloped bends were

limestone," she said. "I shouldn't be able to tell you that without running the conodonts, but I know."

"What if you're wrong?"

"Then I'm wrong, aren't I? They pay me to do the best I can. Geologists are detectives. You work with what you have."

She swung full force with her sledgehammer. The stone did not crack. "This profession is very physical," she complained, and belted the outcrop again.

Her knees sometimes turn black-and-blue when she carries samples down from mountains. She once handed a suitcase to a Greyhound bus driver who said, "What have you got in here, baby—rocks?" She was content to have them ride in the baggage compartment. A geologist I know in California would be unnerved by that. When he travels home from far parts of the world, he buys two airline seats—one for himself and one for his rocks.

We passed Limestoneville. We crossed Limestone Run and the West Branch of the Susquehanna, and now the road was running in a deep crease, a V with sides of about twelve hundred vertical feet: White Deer Ridge and Nittany Mountain—quartzites of early Silurian age, shed west from the Taconic Orogeny. There were quartzite boulders all through the steep woods but a notable absence of outcrops, of roadcuts, of exposures of any kind. In fact, with the exception of the limestones she had collected, we were not seeing much rock to write home about, and Anita was becoming impatient. "No wonder I never did geology in this part of Pennsylvania," she said. "There are no exposures—just colluvium lying in the woods." Multiple ridges were squeezed in close here. Characteristically, the interstate would yield to the country, to the southwestward sweep of the corrugated mountains, as it ran in a valley under a flanking ridge, biding its time for a gap. One would soon appear—not a national landmark with a history of landscape painters and lovestruck Indians, but a water gap, nonetheless—sliced clean through the ridge. Like a fullback finding a hole in the line, the road would cut right and go through. On the far side, it would break into the clear again, veering southwestward in another valley, gradually moving over toward the next long ridge. There would be another gap. Small streams had cut countless gaps. All within twenty miles of one another, for example, were Bear Gap in Buffalo Mountain, Green Gap in Nittany Mountain, Fryingpan Gap in Naked Mountain, Fourth Gap in White Deer Ridge, Third Gap, Second Gap, First Gap, Schwenks Gap, Spruce Gap, Stony Gap, Lyman Gap, Black Gap, McMurrin Gap, Frederick Gap, Bull Run Gap, and Glen Cabin Gap—among others.

In Precambrian, Cambrian, and much of Ordovician time, rivers ran southeastward off the American continent into the Iapetan ocean. Then the continental shelf bent low, and the Martinsburg muds poured into the depression from the east.

Whether they were coming from Africa, Europe, or some accretionary, displaced, hapless Taiwan is completely unestablished, but what is not unestablished is the evidence preserved in the sediment—sand waves, ripple marks, crossbedded point bars—showing currents that flowed west and northwest. In later rock, such evidence is everywhere, showing eastern American rivers flowing toward what is now the middle of the continent all through the rest of Paleozoic time. As each successive orogeny produced another uplift in the east, fresh rivers would pour from it, building their depositional wedges, their minor and major deltas, but running always in a westerly direction. The last orogeny was pretty much spent about two hundred and fifty million years ago, in the middle Permian. For some tens of millions of years after that, the mountains were reduced by weather in a tectonically quiet world. Then, in early Mesozoic time, "earth forces" began to pull the terrain apart. According to present theory, the actual split, deep enough to admit seawater, came at some point in the Jurassic. The Atlantic began to open. On the American side of the break, extremely short steep rivers flowed into the new sea, but for the most part the drainages of what is now the eastern seaboard continued to flow west. By Cretaceous time, the currents had reversed, assuming the present direction of the Penobscot, the Connecticut, the Hudson, the Delaware, the Susquehanna, the Potomac, the James.

Rivers come and go. They are younger by far than the rock on which they run. They wander all over their valleys and sometimes jump out. They reverse themselves and occasionally disappear—their behavior differentiated by textures in the solid earth below. The tightly folded Appalachians are something like the ribs of a washboard. The direction of the structure lies across the direction of scrubbing. In the Paleozoic era, when the tectonic washboard was made and repeatedly lifted from the east, falling rainwater, gathering in streams, found its way westward across the ribs.

With the coming of the Atlantic—the Mesozoic split—the principal drainages of the American East at first continued to flow toward the Midwest. A part of the plate-tectonic story is that a great deal of heat accompanies tectonic rifting and the heat lifts the two sides of the rift like trapdoors facing each other. The shores of the Red Sea look like that. On both sides are mountains, nine, ten, twelve thousand feet high. Extremely short steep rivers fall into the Red Sea. Principal drainages—the intermittent rivers of Arabia—run eastward almost from the east shore many hundreds of miles, and from near the west shore Egyptian rivers run west to the Nile. The world's mid-ocean ridges—the spreading centers of plate tectonics—are configured like the rift of the Red Sea. Typically, the two sides are of gentle pitch, and gradually rise six thousand feet higher than the flanking abyssal plains. Groovelike down the ridgelines run submarine rift valleys. Into the rift valleys of eastern Africa pour extremely short steep rivers, while long ones, like the Congo, rise close to the rift but flow away westward a thousand miles to the sea. It was the discovery and confirmation of spreading centers that opened the story of plate tectonics—and this is still the aspect of the theory that provokes the least debate. Eastern America, in Jurassic time, gradually subsided. The present explanation would be that as the ocean grew wider and the heat of the spreading center became more distant, the region cooled like a collapsing soufflé, while the weight of water and accumulating sediments also pressed down on the continental shelf. In any case, the broad package of land that had tilted northwestward for approximately three hundred million years now seesawed and began again to tilt the other way.

Rivers turned around, pooled temporarily against the ribs of the washboard, and ran over them, seeking weaknesses in the rock. Anew, the running water began to etch out the country. It was a process analogous to photo-engraving, wherein acid differentially eats pictures into treated sheets of metal. The new and reversed eastern rivers differentially eroded the Appalachian structures. Where they got into the shales and the carbonates, they dug deep and wide. Where they found quartzite and other metamorphic rock, they encountered tough resistance. Sometimes, working down into the country, they came to the arching quonset roofs of anticlines, and slicing their way through quartzite found limestones within. It was like slicing into the foil around a potato and finding the soft interior. The water would remove the top of the arch, dig a valley far down inside, and leave quartzite stubs to either side as ridges flanking the carbonate valley. Streams erode headward. That is, they dig back into the valleys they make, eating up the hillsides back into the mountainsides, digging out their grooves up toward the nearest divide. On the other side is another stream, doing the same. Working into the mountain, the two streams struggle toward each other until the divide between them breaks down and they become confluent. One yields to the other—gives up its direction of flow and goes the opposite way, captured. In this manner, some thousands of streams—consequent streams, pirate streams, beheaded streams, defeated streams— formed and re-formed, shifting valleys, making hundreds of water gaps with the general and simple objective of finding in the newly tilted landscape the shortest possible journey to the sea. A gap abandoned by its streams is called a wind gap. In the regional context, the water gap of the Delaware River is a little less phenomenal than it once appeared to be.

Until ten or fifteen years ago, the picture in vogue of the early Cenozoic American East was of a vast peneplain, a flat world of scant relief, with oxbowed meandering rivers heading almost nowhere. The assault of water on the ancestral mountains was thought to have worn down the whole topography close to sea level. The peneplain then rose up, according to the hypothesis, and rivers dissected it, flushing out the soft rock and leaving hard rock high, in the form of remarkably level ridges—as flat as the peneplain, of which they were thought to be remnants. Where the rivers of the peneplain had flowed across the tops of buried ridges, they cut down through them as the ridges came up—making gaps. That was the history as it was taught for three-quarters of a century. It was known as the hypothesis of the Schooley Peneplain, after Schooley Mountain, in New Jersey, which looks like an aircraft carrier. The Schooley Peneplain is out of vogue. It is an emeritus idea. It has been replaced by a story out of steady-state physics having to do with the relationship of level ridgelines to certain degrees of slope. A graduate student once remarked to me that old hypotheses never really die. He said they're like dormant volcanoes.

·   ·   ·

Under the carbonate valleys and quilted farms, the rock was buried from view. The beauty of the fields against steep-rising forests, the shimmer of April green, was not doing much for Anita. She was in need of a lithic fix. Her fingers tapped the wheel. She reminded me of a white-water fanatic on a meandering stretch of flat river. "No wonder I never did

[157]

John McPhee

geology in this part of Pennsylvania," she said again.
It had been a long time between rocks. "I'd really
like to go to Iran someday," she went on, desper-
ately. "The Zagros Mountains are another classic fold-
and-thrust belt. The thing about the Zagros is that
there's no vegetation. You can see everything.
They're a hundred per cent outcrop."

She had scarce uttered the words when the road
jumped to the right and through a nameless gap and
past a roadcut twenty metres high—Bald Eagle
quartzite—and then more and higher roadcuts of
Juniata sandstone in red laminations dipping steeply
to the west. "I take it back. This is one hell of a
series, let me tell you," Anita said. More rock followed, rock in the median, rock right and left, and
we ran on to scout it, to take it in whole. The road
was descending now through gorges of red rock—the
results of precision blasting, of instant geomorphol-
ogy. Their depth increased. They shadowed the
road. And in their final bend was the revealed interior of a mountain, geographically known as Big
Mountain. There had been a natural gap, but it had
not been large enough, and dynamite had contributed three hundred thousand years of erosion.
The entire mountain had been cut through—not just
a toe or a spur. "Holy Toledo! Look at that son of a
bitch!" Anita cried out. "It's a hell of an exposure, a
hell of a cut." More than two hundred and fifty feet
high and as red as wine, it proved to be the largest,
most spectacular man-made exposure of rock on In-

[158]

terstate 80 between New York and San Francisco. It was an accomplishment that might impress the Chinese geological survey. "When you're doing geology, look for the unexpected," Anita instructed me, forgetting the Zagros Mountains.

We stopped on the shoulder in the shadow of the rock. "Holy Toledo, look at that son of a bitch," Anita repeated, with her head thrown back. "Mamma mia!" The bedding was aslant in long upsweeping lines, of which a few were green. Almost due south of Lock Haven and thirty-one miles west of the Susquehanna River, it was Juniata sandstone, brought down off the Taconic uplift and spread to the west by the same system of rivers that transported the rock of the Delaware Water Gap. "This would be a beautiful place to measure the thickness of the section," Anita said. "It's completely exposed. It's consistent. There are no faults. The thin green bands are where deposition was too rapid for oxidation to take place." Evidence of geologists was everywhere. They had painted numbers and letters on the rock. They had removed countless paleomagnetic plugs. "Without these roadcuts, all they can do is drill holes in the ground or find natural streamcuts, which are few and far between," Anita said. The bedding, seen close, was not monotonously even, as rock would be that formed in still water. Instead, it was full of the migrating channels, feathery crossbedding, natural levees, and overbank deposits of its thoroughly commemorated river. There were little maroon mud flakes like raisins. They were plucked off flats in a storm.

We went back a few miles and slowly reviewed the rock. When again we approached the huge road-cut, Anita said, "In Illinois, this would be a state park."

.   .   .

The bedding planes of the Holy Toledo cut, as I would ever after refer to those enormous walls of red stone, were dipping to the east. Over the past few miles, the rock of the country had been folded ninety degrees. To the immediate west, therefore, we would be going down in time and predictably would descend in space to a Cambro-Ordovician carbonate valley, which is what happened, as the road fell away bending left and down into Nittany Valley, where ribs of dolomite protruded here and there among rich-looking pastures turning green, gentle stream courses, white farms. "Penn State sits on Nittany dolomite," Anita said. "It's twenty miles down this valley."

Some remnant Cambrian sandstone formed a blister in the valley. The interstate drifted around it in a westerly way and toward the foot of still another endless mountain—Bald Eagle, the last ridge of the deformed Appalachians. After the Cambrian sandstone, the Ordovician dolomite, there was Silurian quartzite in the gap that broke through the mountain. Its strata dipped steeply west. The rock had bent again, and again we were moving upward through history. Now, though, the dip of the strata would reverse no more. In a dozen miles of ever younger rock, we climbed through the Paleozoic era almost from beginning to end. We went up through time at least three hundred million years and up through the country more than a thousand vertical feet, the last ten miles uphill all the way, from Bald Eagle Creek to Snow Shoe, Pennsylvania—the longest steady grade on I-80 east of Utah—while light, wind-driven snow began to fall.

We had come to the end of the physiographic province of the folded-and-faulted mountains, and the long ascent recapitulated Paleozoic history from the clean sands of the pre-tectonic sea to the dense twilight of Carboniferous swamps. We came up through the debris of three cordilleras, through repetitive sandstones and paper shales—Silurian paper shales, Devonian paper shales, Mississippian paper shales—crumbling on their shelves like acid-paper books in libraries. The shales were so incompetent that they would long since have avalanched and buried the highway had they not been benched —terraced in the manner of Machu Picchu. In other roadcuts, Catskill Delta sandstones, beet-red and competent, were sheer. We had gone through enough hard ridges and soft valleys for me not just to sense but to see the Paleozoic pageant repeatedly played in the rock. For all the great deformity and complexity, the mountains now gone had left patterns behind. The land rising and falling, the sea receding and transgressing, the ancestral rivers losing power through time had not just obliterated much of what went before but had always imposed new scenes, and while I, for one, could not hold so many hundreds of pictures well related in my mind I felt assured beyond doubt that we were moving through more than chaos.

The strata at the foot of the ten-mile hill had been nearly vertical. Gradually, through the long climb, they levelled out. They leaned backward, relaxed, one degree every two million years, until in the end they were flat—at which moment the interstate left the deformed Appalachians and itself became level on the Allegheny Plateau.

The rock was now Pennsylvanian—massive river sandstones of Pennsylvanian time. Flat, decklike, it was comparatively undisturbed. It had been shed, to be sure, from eastern mountains, but had not been much affected by their compressive drive. Crazed streams had disassembled the plateau, leaving half-eaten wedding cakes, failed pyramids, oddly polygonal hair-covered hills. Pittsburgh was built upon such geometries, its streets and roads faithful to the schizophrenic streams, its hills separating its people into socio-racial ethno-religious piles—up this one the snobs, up that the Jews, up this the tired, up that the poor.

A hundred miles northeast of Pittsburgh in the

flurrying snow there were numerous roadcuts now, and in them were upward-fining sequences of sandstones, siltstones, shales—Allegheny black shales—underlying more levels of sandstone, siltstone, and shale. "If you were a prospector for coal, you'd go bananas when you saw these black shales," Anita said. "There ought to be coal in these roadcuts. This is Pennsylvania in the Pennsylvanian—the home office of the rock."

Pennsylvania in the Pennsylvanian was jungle—a few degrees from the equator, like southern Indonesia and Guadalcanal. Freshwater swamp forests stood beside the nervously changing coastline of a saltwater bay, just as Sumatran swamps now stand beside the Straits of Malacca, and Bornean everglades beside the Java Sea. In Pennsylvanian time, glacial cycles elsewhere in the world were causing sea level to oscillate with geologic rapidity. The swamps pursued the shoreline as the sea went down. Marine limestone buried the swamps as the sea returned. In just one of these cycles, the shoreline would move as much as five hundred miles—the sea transgressing and regressing through most of Pennsylvania and Ohio. There were so many such cycles at close intervals in Pennsylvanian time that Pennsylvanian rock sequences are often striped like regimental ties—the signature of glaciers half the world away. They existed three hundred million years ago, and glacial patterns of that kind have not been repeated until now, when the measure of our own brief visit to the earth is being recorded as a paper-thin stripe in Pleistocene time.

On both sides of the interstate, above the silhouettes of screening trees, we saw the tops of draglines—the necks and heads of industrial giraffes. They and predecessor machines had been working for fifty years, altering the topography, stripping the coal beds of Pennsylvania—in all, a mineral deposit worth a great deal more than the diamond mines of Kimberley and the goldfields of the Klondike. Coal was in the roadcuts now and would continue to be for many tens of miles—in layers that were not the dull deep gray of the Allegheny shale but truly black and shining. Layered light and dark, the roadcuts looked like Hungarian tortes. Reading up from the bottom, there was sandstone, siltstone, shale, coal, sandstone, siltstone, shale, coal. We would see limestones farther on, capping the coal where sea had covered the swamps. The present sequence was built behind a coastline—as is happening now, for example, in the bayous of the Mississippi Delta—by rivers meandering to and fro, covering with sand the matted vegetation. "These roadcuts are a textbook on the making of coal," Anita said. Buried and compressed, vegetal debris first becomes peat—a mélange of spores, seed coats, wood, bark, leaves, and roots which looks like chewing tobacco and burns about as well. Peat bears much the same relation to coal that snow does to glacier ice. As snow is ever more buried and compacted, it recrystallizes and becomes ice—on the average ten times as dense as the original snow. As peat is buried, compacted, subjected to geothermal heat, it gradually gives up much of its oxygen, hydrogen, and nitrogen, and concentrates its content of carbon. The American Geological Institute's "Glossary of Geology" defines coal as "a readily combustible rock." By weight, any rock that is half carbonaceous material is coal. Its density is roughly ten times the density of peat. In the United States, there is enough peat to keep Ireland warm for a thousand years. The United States uses almost none of it, because the United States also happens to have a great deal more coal than any other country in the world, with the exception of the Soviet Union, which has a like amount. Peat that remains near the surface will never become coal. Buried three-quarters of a mile, it becomes bituminous. With a microscope, you can see wood and bark, leaves and roots, seed coats and spores in bituminous coal—and even identify the plants they came from. Buried deeper and folded severely under pressure, it becomes anthracite. Anthracite is roughly ninety-five per cent carbon and is so hard that it fractures conchoidally, like an arrowhead. Anthracite is iridescent, and burns with a clear blue flame. Coal is a record of tectonics. In late Pennsylvanian time, when the third set of mountains came up in the east and shed still another wedge of debris, kneading it into what had gone before, the great pressure, deep burial, and severe folding produced the anthracites of eastern Pennsylvania, the pod-shaped coalfields of the folded-and-faulted mountains, which erosion and isostasy have lifted from the depths. Anthracite seams are often upside down or standing on end. Here in the Allegheny Plateau, burial was reasonably deep but tectonic pressures were minor, and the result is a lesser grade of coal.

We stopped and tried to collect some but had difficulty finding a sample that would not break up in the hand. "This is very flaky, high-ash coal," Anita said. "People take it anyway. They come out to these roadcuts with buckets and take it home to burn."

We moved on through miles of coal-streaked roadcuts, and topographically to somewhat higher ground, where the coal seams were thicker. "As you go westward and upsection, you get more coal, because the rivers, growing older, became more sluggish," Anita said. "The floodplains became broader. There was more ponded water. There was more area for vegetation to grow and accumulate—like the lower Mississippi Valley today." About five miles east of Clearfield, we stopped at a long, high throughcut full of coal. Draglines were working on both sides of the road. We chipped out some samples with rock hammers. The samples had integrity. "This is a hell of a coal," Anita remarked. "Good commercial coal. To make it, there would have been about three thousand feet of Pennsylvanian stuff on top of it, which has been removed by erosion. Three thousand feet is the amount of overburden that will produce coal of this rank."

Stirred within by all these free B.t.u.s (twelve thousand per pound), I flailed at the cut with my rock hammer and filled a bag with good commercial coal, to take home and burn in my stove. Anita commented that coal dust was blacking my face.

I wiped at it with a bandanna, and asked her, "Did I get it all?"

She said, "Good enough for government work." And we headed up the road.

.   .   .

When the final great pulse of mountain-building folded eastern Pennsylvania, the deep burial and tectonic crush may have done wonders for the coal seams there, but all the oil in the country rock was burned black and destroyed. Conodonts were blackened, too. As Anita's many samplings would prove, conodonts become lighter in color and hue in a westward trend across the state—from black to cordovan to dusky orange to brightening levels of yellow. Running west of Du Bois and Clarion now, and less than fifty miles from Ohio, we were out of the browns and well into the gold. If the quality of coal improves eastward, the theoretical quality of petroleum goes the other way. We took a right off the interstate. Soon we were cruising on Petroleum Street, in downtown Oil City.

We continued north. In the fifteen miles between Oil City and Titusville lay the Napa Valley of early American oil. It was a V-shaped, intimate valley, five hundred feet from rim to river, and along its floor were oil refineries so small they were almost cute. They did not suggest the starry lighted skeletal cities of Exxon's Bayway Refinery or Sunoco's Marcus Hook. They suggested Christian Brothers, the Beringer Winery, the Beaulieu Vineyard. One refinery followed another. Wolf's Head, Pennzoil. They stood beside Oil Creek, which was so named in the eighteenth century because petroleum dripped out of its banks and into the water. Indians had found it, three centuries before, to judge by the age of trees that were growing in pits they had dug to collect the oil in pools. The Senecas rubbed their skins with it. They may have used it for light and heat. The use of petroleum is old in the world. Workmen laid asphalt three thousand years before Jesus Christ. The first energy crisis involving petroleum was in 1875 B.C. The first oil spills were natural, and were not so large that they could not be cleaned up by bacteria that feed on oil. In 1853, in California, a lieutenant in the Corps of Engineers reported that "the channel between Santa Barbara and the islands is sometimes covered with a film of mineral oil, giving to the surface the beautiful prismatic hues that are produced when oil is poured on water." Always, it was found in seeps. Even until a few years after the Second World War, all Iranian oil fields were associated with surface seeps. The first well in Texas—1865—was drilled near a seep. A well in Ontario had been drilled six years earlier, and in the same summer the first commercial oil well in the United States was drilled in Pennsylvania by Colonel Edwin Drake—less than a hundred steps from Oil Creek.

Colonel Drake had no record of military service. He was a sick railroad conductor—forty years old but debilitated, too fragile to remain upright in the lurching aisles of the New York & New Haven. To the Pennsylvania Rock Oil Company of Connecticut, which had bought farmland and timberland along Oil Creek, he had committed his life savings. Drake was not a geologist. He did not know that petroleum is primarily the remains of marine algae that pile up dead on the floors of shallow seas in situations that prevent oxidation. He did not know that the algal corpses slowly stew for millions of years at temperatures just high enough to crack them into crude. He did not know that oil forms in one kind of rock and moves into another—forms in, say, the lagoonal muds of epicontinental seas, and moves later into the sandstones that were once the barrier beaches between the lagoons and the open sea. He did not know that Oil Creek had cut down through Pennsylvanian and Mississippian formations and on into a Devonian coast. Drake knew none of this in 1859, and neither did the science of geology. What Drake did know was that there was negotiability in the stuff that was dripping into Oil Creek. It was even used as medicine. Fleets of red wagons had been all over eastern America selling seepage as a health-enhancing drink. "Kier's Genuine Petroleum! Or Rock Oil! A natural remedy . . . possessing wonderful curative powers in diseases of the Chest, Windpipe and Lungs, also for the care of Diarrhea, Cholera, Piles, Rheumatism, Gout, Asthma, Bronchitis, Scrofula, or King's Evil; Burns and Scalds, Neuralgia, Tetter, Ringworm, obstinate eruptions of the skin, Blotches and Pimples on the face, biles, deafness, chronic sore eyes, erysipelas . . ." Drake had, in addition, the encouragement of a Yale professor of chemistry who ran a bottle of the seepage through his lab and said, "It appears to me . . . that your company have in their possession a raw material from which, by simple and not expensive process, they may manufacture very valuable products. It is worthy of note that my experiments prove that nearly the *whole* of the raw product may be manufactured without waste." And what Drake had, above all, was the inspiration to go after the substance in its reservoir rock, not to be content to blot it up from the streambanks but to drill for it, never mind that he was making a fool of himself in the eyes of the local rubes. He would punch their tickets later. At sixty-nine and a half feet, he completed his discovery well.

There was an oil rush to Oil Creek, and frontier conditions in shantytowns, and forests of derricks on denuded hills. There was a town called Red Hot, Pennsylvania. There was Petroleum Centre. Pithole City. Babylon. In three months, the population of Pithole City went from nobody to fifteen thousand. River flatboats carried the oil to market. Their holds

were divided into compartments, much as the holds
of supertankers are divided now. Millers in the valley
were paid royalties to release water on cue from
millponds, raising the level of the creek to float the
flatboats downstream. They sometimes broke and
spilled.

The Dramatic Oil Company was established in
the valley by John Wilkes Booth, who ruined his well
trying to make it more productive. With failure, he
departed, in the fall of 1864, to look for other things
to do.

I am indebted for many of these facts to Ernest
C. Miller, of the West Penn Oil Company, who col-
lected them for the Pennsylvania Historical and
Museum Commission.

By 1871, oil was being pumped from the ground
in nine countries, but ninety-one per cent of world
production still came from Pennsylvania. When it
was distilled into its components—paraffin, kero-
sene, and so forth—the gasoline, which in those days
had no commercial value, was poured off into the
ground.

Petroleum is rare because it represents an extremely low percentage of the life that has lived on
earth. In rock, the ratio of all organic carbon to
petroleum carbon is eleven thousand to one. For
petroleum carbon to turn into oil and be preserved,
many conditions have to align, the most important of
which is the thermal history of the source rock—the
temperature through time as recorded by, among

other things, the colors of conodonts. "The petroleum in this valley makes some of the best lubricating oil in the world," Anita said. "It is a very low-specific-gravity oil and needs little refining, because it has been refined to near-perfection by natural earth processes. It has been at low temperatures—around a hundred degrees Celsius—for maybe two hundred million years. You can practically take it out of the ground and put it in your car."

For a hundred and fifty miles, we had been traversing country that was free of glacial drift. Nowhere to be seen were the tills and erratics, the drumlins and kames left behind by Wisconsinan ice. Like a lifted hem, the line of maximum advance had been up in New York State somewhere, but now, in westernmost Pennsylvania, the glacial front had billowed south, and where Interstate 80 meets the 80th meridian we again crossed the terminal moraine. Sign of the ice was everywhere—the alien boulders in the woods, the directional scratches on the country rock, the unsorted gravels, cobbles, and sands. The signature of glaciation is as bold as John Hancock's and as consistently recognizable wherever ice has moved across the solid earth. In the presence of the evidence, one has no difficulty imagining the arctic ambience, the high blue-white ice lobes thickening to the north, the white surface wide as the continent and swept by uninterrupted gales, the view in sunlight blinding, relieved only by isolated mountain summits, ice moving around them in the way that water slides past boulders in a stream.

Welcome to Ohio. A sign in the median said "STAY AWAKE! STAY ALIVE!" Ohio is not rich in road-cuts. It is a little less poor, however, than Indiana, Illinois, Iowa, and Nebraska, and before long we were running through burrowed marine shales and walls of lithified river sand. It was rippled Carboniferous sandstone. We were still in rock of that age, but gradually and imperceptibly we had been losing altitude since we climbed the Allegheny front. The eastern rim of the plateau had been more than two thousand feet high, and by now we were down to half that, as we moved farther away from the ancestral mountains and their wedge of sediment thinned.

We had come into the continent's province of supreme tectonic calm, the Stable Interior Craton, where a thin veneer of sediment lies flat upon the stolid fundament, where the geology—even by geological standards—is exceptionally slow. "This is the most conservative part of the U.S.," Anita said. "I've often thought about it. The wildest, craziest people are in the most tectonically active places."

And yet the craton stirs. There is no part of the face of the earth that vertically and laterally does not move. The bedding planes in Midwestern rock, which appear to be absolutely level, do in fact dip. They will descend across a great many miles and then rise, arching over the far rim of a vast and shallow bowl, and then subtly dip again to form a similar bowl: the Cincinnati Arch, the Michigan Basin, the Kankakee Arch, the Illinois Basin. Anita called the arches "basement highs." She said Hudson Bay is a continental basin, slowly filling up. The basins of the Midwest are filled to the brim with level ground. They are products of the creaking motions of the craton, in response perhaps to plays of force from deep within the mantle—a process that, in the general phrase, is "not well understood." They represent a degree of tectonic activity about as lively as the setting in of rigor mortis. This has not always been the regional story. There are roots of long-gone mountains deep in the rock of the stable craton, but it has not had an orogeny in a thousand million years. "What has the Midwest been doing since then? It's been sitting around doing nothing," Anita said. "It has just sat here ho-humming." Shallow seas may have quietly arrived and departed, and coal beds formed in the ground, but in all that time there has been no occurrence that can begin to rival in scope or total change the advent from the north of walls of marching ice.

The ice was Antarctic in breadth. The traceable episodes of recent continental glaciation have each placed about as much ice over North America as is upon Antarctica now. In Wisconsinan time, which lasted about seventy-five thousand years and ended ten thousand years ago, three-fifths of all the ice in the world was on North America, another fifth covered much of Europe, and the rest was scattered. Of all special fields within the science, glacial geology is the most evident, the least inferred. It is, for one thing, contemporary. The ice is in recess but has not gone away. In addition to the ice of Antarctica, there is ice more than two miles thick over Greenland. There are twenty-seven thousand square miles of ice on Alaska (four per cent of Alaska). In Alaska, as in Switzerland and elsewhere in the world, you can see cirque glaciers feeding into the master glaciers of alpine valleys. You can see that the cirque glaciers have dug scallops into the high ridges, and where three or four cirque glaciers have been arranged like petals they have torn away the rock until all that remains is a slender horn—the Kitzsteinhorn, the Finsteraarhorn, the Matterhorn. Not only are ice sheets, ice fields, and individual glaciers operating today with effects observable as motions occur, but wherever they once flowed their products remain in abundance and intact. They have come and gone so recently.

The evidence may seem obvious now, but not until the eighteen-thirties did mankind comprehend its significance. There had been insights, hints, and clues. James Hutton, the figure from the Scottish Enlightenment who by himself developed the novel view of the world on which modern geology rests, mentioned in his "Theory of the Earth" (1795) that the gravels and boulders of Switzerland's great valley appeared to have been put there by ancient extensions of alpine ice. But Hutton, who formed his theory among the scratched granites and drifted gravels of Scotland, never suspected that Scotland itself had been a hundred per cent covered—actually dunked into the mantle—by ten thousand feet of ice.

In 1815, in the Swiss Val de Bagnes, below the Pennine Alps, a mountaineer remarked to a geologist that all those big boulders standing around in odd places had been carried there by a glacier long since gone. The mountaineer's name was Perraudin. He was a hunter of chamois. The geologist was Jean de Charpentier. He did not believe the hunter and ignored the information. In Europe, Noah's Flood had for so long been regarded as the principal sculptor of the earth that almost no one was inclined to hazard an alternative interpretation. If boulders were out of touch with bedrock of their type, diluvian torrents had moved them, or flows of diluvian mud. In 1821, a Swiss bridge-and-highway engineer named Ignace Venetz told the Helvetic Society of Natural Sciences that he believed what the mountaineer had told Charpentier. He believed, in addition, that boulders had been scattered all over Switzerland by glaciers of *"hauteur gigantesque"* from *"une époque qui se perd dans les nuits des temps."* Venetz was ignored, too—until Charpentier decided, twelve years later, that his suppositions were probably correct. Char-

twentieth-century field geologist when he set out with Charpentier to stroll through the valley of the upper Rhone. What Agassiz saw forever altered his life, as ice had altered the valley. When he left, he had no remaining doubt of the truth of what Perraudin, Venetz, and Charpentier believed. Wandering the Swiss countryside low and high, he found further evidence everywhere he went—grooved rock, polished rock, moraines where ice had long been gone, boulders rounded off and set where water never could have shoved them. He visited similar landscapes in enough places to spread far in his imagination the contiguity they implied, and in one spark of intuition he saw the ice covering more than the valley, the canton, the nation. The idea of continental glaciation fell into place—a stunning moment of realization that ice many thousands of feet thick had been contiguous from Ireland to Russia. When the Helvetic Society met in Neuchâtel in the summer of 1837, Louis Agassiz—as its president-elect—addressed the savants. Instead of reading an expected discourse in paleontology, he outlined at great length the evidence and chronology of glacial history as he had come to see it, announcing to the Society and to the world at large what would before long be known as the Ice Age.

He called it the *Époque Glaciaire*. By any name, at home or abroad, it did not overwhelm his colleagues. He was attacked far more than defended. Von Buch literally threw up his hands, and not without the perspectives of the future partly on his side, for Agassiz—like the "plate-tectonics boys," as seen by Anita Harris—had not known where to stop. His remarks had gone beyond his reconstruction from observable phenomena of a cover of ice across the whole of northern Europe: he had concluded that the newborn Alps, rising under the ice, had caused it to break up.

Agassiz's friend and mentor Alexander von Humboldt, whose name reposes in the western Americas in the Humboldt Current and the Humboldt River, strongly urged Agassiz to go back to cataloguing fossil fishes, the work for which Agassiz was internationally known and for which the Geological Society of London had awarded him the Wollaston Medal. "You spread your intellect over too many subjects at once," he wrote to Agassiz. "I think that you should concentrate . . . on fossil fishes. In so doing, you will render a greater service to positive geology than by these general considerations (a little icy withal) on the revolutions of the primitive world. . . . You will say that this is making you the slave of others; perfectly true, but such is the pleasing position of affairs here below. Have I not been driven for thirty-three years to busy myself with that tiresome America . . . ? Your ice frightens me."

Agassiz's response was to address himself more intensively than ever to glaciers—glaciers of the present and the past. "Since I saw the glaciers I am quite of a snowy humor, and will have the whole surface of the earth covered with ice, and the whole prior creation dead by cold," he wrote in English to an English geologist. "In fact I am quite satisfied that ice must be taken in every complete explanation of the last changes which occurred at the surface of Europe." He found moraines on the plains of France. He found Swedish boulders in Germany. In Grindelwald, a stranger heard his name and, seeing his boyish appearance, asked if he was the son of the great and famous professor.

In 1839, Agassiz went to the glaciers on the apron of the Matterhorn, the glaciers under the Eiger and the Jungfrau. He walked up the Aar Glacier to the base of the Finsteraarhorn, the highest peak in the Bernese Oberland. "There I ascertained the most important fact that I now know concerning the advance of glaciers," he wrote later. From a message in a bottle in a cabin on the ice, he had learned that the monk who built the place in 1827 had returned nine years later to find it more than two thousand feet down the mountain. Agassiz established his own shelter on the Aar Glacier. He and his colleagues drove stakes into the ice—a row of them straight across the glacier—and before long discovered that glacier ice, like a river, flows more rapidly in the center and also tends to speed up toward the outsides of bends. Diverting a meltwater stream that was pouring into a deep hole in the ice, he set up a sturdy tripod at the surface and had himself lowered into the glacier. He was twenty fathoms down in a banded sapphire world when his feet touched water and he shouted instructions that his descent be stopped. His colleagues on the glacier misinterpreted his cry and lowered him into the water. The next shout was different and was clearly understood. The dripping Agassiz was raised toward the surface among stalactites of Damoclean ice, so big that had they broken they would have killed him. Concluding the experiment, he said, "Unless induced by some powerful scientific motive, I should not advise anyone to follow my example." The better to see the alpine-valley ice in its regional perspective, Agassiz and his team climbed mountains—they climbed the Jung-frau, the Schreckhorn, the Finsteraarhorn—and made their observations from the summits, com-pletely unmindful atop a number of the mountains that no one had been there before.

Agassiz went to England, Scotland, Ireland, and Wales, looking for the tracks of glaciers. He found them in England, Scotland, Ireland, and Wales. As in Switzerland, he saw *roches moutonnées*—humps of exposed bedrock that were characteristically smooth on the side from which the ice had arrived, and plucked and shattered on the other. "The surface of Europe, adorned before by a tropical vegetation and inhabited by troops of large elephants, enormous hip-popotami, and gigantic carnivora, was suddenly buried under a vast mantle of ice, covering alike plains, lakes, seas, and plateaus," he wrote in his *"Études sur les Glaciers"* (1840). "Upon the life and movement of a powerful creation fell the silence of death. Springs paused, rivers ceased to flow, the rays of the sun, rising upon this frozen shore (if, indeed, it was reached by them), were met only by the breath of the winter from the north and the thunders of the crevasses as they opened across the surface of this icy sea."

The reception all this got continued to be colder than the ice. Von Buch, author of the first geological map of Germany and already celebrated for his studies of volcanism, did not conceal his indignation. In fact, he had apparently removed Agassiz's name from consideration for a professorial chair at the University of Berlin. Sir Roderick Murchison, the Scottish geologist who had identified and named the Silurian system, warned that he was prepared to "make fight." Addressing the Geological Society of London, he said, "Once grant to Agassiz that his deepest valleys of Switzerland, such as the enormous Lake of Geneva, were formerly filled with snow and ice, and I see no stopping place. From that hypothesis you may proceed to fill the Baltic and the northern seas, cover southern England and half of Germany and Russia with similar icy sheets, on the surfaces of which all the northern boulders might have been shot off. So long as the greater number of the practical geologists of Europe are opposed to the wide extension of a terrestrial glacial theory, there can be little risk that such a doctrine should take too deep a hold of the mind."

Whatever the cause, the effects Agassiz was studying impressed von Humboldt as purely local phenomena. Agassiz's *"descente aux enfers"*—into the innards of the glacier—alarmed his friend as a physical risk commensurate with the risk Agassiz was taking with his paleontological reputation. Von Humboldt wrote to say that he had now "read and compared all that has been written for and against the ice-period" and that he was no closer to accepting the theory. He quoted Mme. de Sévigné's saying that "grace from on high comes slowly." And added, "I especially desire it for the glacial period."

The turnabout was at hand, however. Charles Lyell, the most outstanding British geologist of the nineteenth century, closely read the *"Études sur les Glaciers"* and found himself enlightened. "Lyell has adopted your theory in toto!!!" a friend wrote to Agassiz. "On my showing him a beautiful cluster of moraines, within two miles of his father's house, he instantly accepted it, as solving a host of difficulties that have all his life embarrassed him." Charles Darwin hurried out into the countryside to see for himself if there were "marks left by extinct glaciers." He wrote to a friend, "I assure you, an extinct volcano could hardly leave more evident traces of its activity and vast powers. . . . The valley about here and the site of the inn at which I am now writing must once have been covered by at least eight hundred or a thousand feet in thickness of solid ice! Eleven years ago I spent a whole day in the valley where yesterday everything but the ice of the glaciers was palpably clear to me, and I then saw nothing but plain water and bare rock."

The scientific dons of Cambridge continued stubborn, but—as would happen with the theory of plate tectonics in the years following the revelations of the nineteen-sixties—geologists in expanding numbers accepted the glacial picture, and before long there was a low percentage that did not enthusiastically subscribe. Delivering an address in 1862 to the Geological Society of London, Sir Roderick Murchison declared without shame that he, too, now saw the picture. He sent a copy of his address to Agassiz with a note that said, "I have had the sincerest pleasure in avowing that I was wrong in opposing as I did your grand and original idea of my native mountains. Yes! I am now convinced that glaciers did descend from the mountains to the plains as they do now in Greenland."

. . .

Greenland is eighty-five per cent capped with ice. Anyone who doubts that we live in a glacial epoch need only note the great whiteness that Greenland contributes to a map. "The ice melted here eighteen thousand years ago," Anita said, with a nod toward the roadside in Ohio. "It melted twelve thousand years ago in Wisconsin and Maine. If you ask a penguin in the Antarctic, the Ice Age hasn't stopped yet."

The ice on Antarctica, six million square miles, is also (generally) two miles thick. "You get ice caps when you have landmasses in the polar positions," Anita went on. "The only thing worse would be if the Siberian landmass were sitting over the North Pole. Then, God help us, things would be really bad. As it is, the sea ice at the North Pole is only six feet thick. It takes a continent to support a really heavy sheet of ice. If the ice of Greenland and Antarctica were to melt now, sea level would go up at least a hundred feet. Think what the water would cover. Half the cities in the world. In the South, you can be three hundred miles from the coast and only fifty feet above sea level. Through most of time, the earth has been without ice caps. Twenty thousand years ago, when there was much more ice than there is now, the sea was three hundred feet lower. The coast was more than a hundred miles east of New York. You could have walked to the edge of the continental shelf. Baltimore Canyon, Hudson Canyon were exposed in the open air."

Outside the automobile window were three landscapes, trifocal, occupying separate levels in time and mind. Latently pictured in the rock beside the road was the epicratonic sea of three hundred and twenty million years ago, with the Cincinnatia Islands off to the west somewhere, in what is known to geologists as Ohio Bay. There was also evidence of the deep ice of twenty thousand years ago, with its lobate front some distance to the south, near Canton,

Massillon, and Wooster. And there was, of course, the slightly rumpled surface of the modern state of Ohio, looking like a bedspread on which someone had taken a nap. Not nearly as flat as the rock below was the undulating interstate, where diesel exhausts were pluming and Winnebagos were yawing in the wind.

"The goal of many geologists is to make time-lapse maps of earth history," Anita remarked. "Look at topographic maps from just a hundred years ago for coastal areas of low relief, and the changes are tremendous."

We went through a ten-metre roadcut of massive sandstone so rich in iron it had rusted the road. Being tough by comparison with its neighboring rock, it stood high and formed a hill, and hence it had been blasted to convenience the interstate. "That is one hell of a sandstone," Anita said with enthusiasm, seeing in it something I could not discern.

We crossed a river. "That was the well-known Cuyahoga," she said. "If you swim in it, you dissolve."

The Cuyahoga was flowing south. It rises in northeasternmost Ohio, runs south into Akron, then reverses its direction, swinging north through Cleveland and into Lake Erie.

More warning signs flashed by. "STAY AWAKE! STAY ALIVE!"

Anita said, "I'm trying. I'm trying."

Now spanning the road was an Italianate steel-arch bridge, standing on Berea sandstone, a fragment of the Berea Delta, of early Mississippian age, which had extended its bird-foot shape far into Ohio Bay. We stopped, and picked quartz pebbles the size of golf balls out of a conglomerate there. "These would have been just offshore," she said. "You can take the pebbles out of the rock with your hands because it was never heated up like the conglomerate at the Delaware Water Gap. This was never buried much. It is not well lithified. It hasn't experienced enough heat to get tough."

A few miles west, we crossed the Cuyahoga River again, and looked down some distance from the interstate bridge into the Cuyahoga's extensively reamed-out valley, with its modest, meandering stream.

"It's an underfit stream," said Anita. "A little half-ass stream in a valley made wide by glacier ice. The Cuyahoga's valley was steepened and en-trenched, like Yosemite."

"You are comparing the Cuyahoga Valley with Yosemite?"

"Technically."

We left the interstate and followed the valley into Cleveland. The Cuyahoga River has suffered a bad press. When it caught fire a few years ago, it attracted national attention. Its percentage of water had become low relative to its content of hydrogen in various combinations with carbon. The river

burned so fiercely that two railroad bridges were
nearly destroyed. There was no mention in the papers of the good things the river had done. It had
made parks. It had been there before the glacier ice
and had cut down five hundred feet through Mississippian formations into Senecan and Chautauquan
time—stages of the late Devonian. It cut deep ra-
vines, which the ice later broadened into canyons.
The ice augered through the V-shaped valley and
turned it into a U. Which is what ice did at Yosemite
—with the difference that the walls of Yosemite are
speckled white granite, while the canyon walls of
Cleveland are flaky black gasiferous anoxic shale. As
mud, the shale was deposited in quiet water in a late
Devonian sea. The rock contains the unoxidized remains of so many living things that it is by volume as
much as twenty per cent organic. In thin laminations,
it grew layer upon layer—paper shale. "The water
was so quiet you can trace the same little lens forever," Anita said. "The formation produces gas like
crazy. The gas migrates up into the sandstone above,
which holds it. Berea sandstone. People drill their
own wells to the Berea and heat their homes." Much
of Cleveland's metropolitan-park system is in the
deep Yosemite of the Cuyahoga River, under paper-
flake carbon cliffs—a natural world of natural gas.

Like the Cuyahoga today, most rivers in Ohio
before the recent ice sheets looked for outlets to the
north and northwest. Nearly all were wiped away by
the planing drive of ice. Water pooled against the

glacial front and spilled away to the south and west. It skirted the ice, roughly tracing its southernmost outline, forming a new river system and a "periglacial valley"—the Ohio River, the Ohio Valley.

.   .   .

When Darwin published "The Origin of Species," its affront to organized religion did not altogether exceed the dismay that was felt in science. Even Sir Charles Lyell said, "Darwin goes too far." Thomas Henry Huxley and a few others were supportive, but almost every paleontologist in the British Isles was flat negative, and the geologist Adam Sedgwick, of Cambridge University, who, with Murchison, had discerned and established the Devonian system, described himself reading Darwin "with more pain than pleasure." He said, "Parts of it I admired greatly, parts I laughed at till my sides were almost sore; other parts I read with absolute sorrow, because I think them utterly false and grievously mischievous. Many . . . wide conclusions are based upon assumptions which can neither be proved nor disproved. . . . Darwin has deserted utterly the inductive track and taken the broadway of hypothesis." Applause for Darwin was even sparer from scientists across the Channel, with the notable exception of the Belgian geologist J. J. d'Omalius d'Halloy, who, as it happened, had subscribed from the beginning to Louis Agassiz's glacial theory as well, and whose *"Terrain Crétacé"* was the discovery ground for the worldwide Cretaceous system.

In the United States, by contrast with Europe, geologists, biologists—the scientific community at large—were for the most part quick supporters and early participants in the sweep of evolution. In the United States, also, there was a notable exception. He was Professor Louis Agassiz, of Harvard University. He had crossed the Atlantic and given a few lectures. He stayed for the rest of his life. He became, as he has remained, one of the most celebrated professors in the history of American education. It was a renown that rested largely on his amazing and infectious capacity for talking about ice. Never mind that he could not speak schoolroom English. His words drew pictures of glaciers in motion, many thousands of feet thick and larger by far than the Sahara. His words drew pictures of glacier ice over Boston, in the act of depositing Cape Cod; of glacier ice over Bridgeport, in the act of depositing Long Island; of ice retreating from Concord, leaving Walden Pond. Harvard was, at core, a drumlin, a glacial coprolite, packed in recessional outwash. America excited Agassiz, as well it might, for it had held the greater part of the ice he had dreamed of, covering the world. He went to Lake Superior and paddled its shoreline in a bark canoe. The features he saw there he had known in Neuchâtel. He went to the Hudson Highlands and remembered the highlands of the Rhine. "The erratic phenomena and the traces of glaciers . . . everywhere cover the surface of the country," he wrote. "Polished rocks, as distinct as possible; moraines continuous over large spaces; stratified drift, as on the borders of the glacier of Grindelwald." He went to the Connecticut Valley: "The erratic phenomena are also very marked in this region; polished rocks everywhere, magnificent furrows on the sandstone and on the basalt, and parallel moraines defining themselves like ramparts upon the plain. . . . What a country is this! All along the road between Boston and Springfield are ancient moraines and polished rocks. No one who had seen them upon the track of our present glaciers could hesitate as to the real agency by which all these erratic masses, literally covering the country, have been transported. I have had the pleasure of converting already several of the most distinguished American geologists to my way of thinking."

Henry David Thoreau took Agassiz's book out of the Harvard library and returned it a few weeks later —perhaps unread. Apparently, Thoreau never knew that Walden Pond was a glacial kettle, had no idea that he lived among moraines and drumlins, ice-transported hills. Although he and Agassiz were acquainted and shared the same part of Massachusetts for sixteen years, there is in Thoreau's work no discussion of glaciation. Thoreau evidently never suspected that all his Nausets and Chesuncooks, Merrimacks and Middlesex ponds had been made and shaped by ice.

Agassiz was so caught up in glacial and general

geology that he would try to teach it to stagecoach drivers. He believed that anyone, given a little help, could understand the nature of the earth. In Boston, in order to make his case perfectly and avoid the rockslides of his Franco-Germanic accent and syntax, he announced that he would give a series of lectures in French on the *Époque Glaciaire*. People paid to hear it, and he preserved their admiration in re-crystallized *mots justes*. When he spoke of the Jura, the Pennine Alps, and the boulders in the valleys between, no one was as moved as Agassiz. His great range of expression did not exclude tears. With his large forehead, full lips, aquiline nose, and shoulder-flowing hair, he all but held a baton in his hand with which to conduct the movements of the ice. One Saturday a month, he met with his friends for a late, seven-course lunch from which no one was in a hurry to go home. They would meet, like as not, at "Parker's" in Boston, in a room looking out on City Hall. "Agassiz always sat at the head of the table by native right of his large good-fellowship and intense enjoy-ment of the scene," his friend Sam Ward eventually recalled. Henry Wadsworth Longfellow generally sat at the other end, with Oliver Wendell Holmes on his right. Holmes preferred his windowlight over the shoulder. On around the table were James Russell Lowell, John Greenleaf Whittier, Nathaniel Haw-thorne, Ralph Waldo Emerson, Richard Henry Dana, Jr., Ebenezer Hoar, Benjamin Peirce, Charles Eliot Norton, and James Elliot Cabot, among others. Agas-siz, with a glass of wine at his elbow, would sometimes conduct the conversation with two lighted cigars, one in each hand. Holmes said of him that he had "the laugh of a big giant." Longfellow was relieved and pleased when Agassiz told him he liked the description of the glacier in "Hyperion." Emerson in his journal described Agassiz as "a broad-featured unctuous man, fat and plenteous." Sir Charles Lyell was invited, on his visits to America. United States Senator Charles Sumner was occasionally present as well. Agassiz was indifferent to him, because Sumner showed too much interest in politics.

The group was known as Agassiz's Club, more officially as the Saturday Club. One summer, when the club went off to the Adirondacks on a camping trip, Longfellow refused to go, because Emerson was taking a gun. "Somebody will be shot," said Longfellow, explaining that Emerson was too vague to be trusted with a gun. Longfellow's works of poetry include a birthday ballad in praise of Agassiz, which Longfellow read aloud at the Saturday Club, and in which Nature addressed the Professor:

> *"Come wander with me," she said,*
> *"Into regions yet untrod;*
> *And read what is still unread*
> *In the manuscripts of God."*

John Greenleaf Whittier also wrote a poem about Agassiz, more than a hundred lines in length, ten of which are these:

Said the Master to the youth:
"We have come in search of truth,
Trying with uncertain key
Door by door of mystery;
We are reaching, through His laws,
To the garment-hem of Cause,
Him, the endless, unbegun,
The unnameable, the One
Light of all our light the Source,
Life of life, and Force of force."

Longfellow, travelling in Europe in 1868, called on Charles Darwin. "What a set of men you have in Cambridge," Darwin said to him. "Both our universities put together cannot furnish the like. Why, there is Agassiz—he counts for three."

Darwin's generosity was remarkable in the light of Agassiz's reaction to "The Origin of Species." As Agassiz summarized it: "The world has arisen in some way or other. How it originated is the great question, and Darwin's theory, like all other attempts to explain the origin of life, is thus far merely conjectural. I believe he has not even made the best conjecture possible in the present state of our knowledge." Agassiz never accepted Darwinian evolution. Many years earlier, as a young man, and as a result of his paleontological researches, he wrote the following: "More than fifteen hundred species of fossil fishes with which I have become acquainted say to me that the species do not pass gradually from one to the other, but appear and disappear suddenly without direct relations with their predecessors; for I do not think that it can be seriously maintained that the numerous types of Cycloids and Ctenoids, which are nearly all contemporaneous with each other, descend from the Placoids and Ganoids. It would be as well to affirm that the mammals, and man with them, descend directly from fishes. All these species have a fixed time for coming and going; their existence is even limited to a determined period. And still they present, as a whole, numerous, and more or less close affinities, a determined coördination in a system of organization which has an intimate relation with the mode of existence of each type, and even of each species. More still: there is an invisible thread which is unwinding itself, through all the ages, in this immense diversity, and offers as a final result a continuous progress in this development of which man is the termination, of which the four classes of vertebrates are the intermediate steps, and the invertebrates the constant accessory. Are not these facts manifestations of a thought as rich as it is powerful, acts of an intelligence as sublime as provident? . . . This is, at least, what my feeble intellect reads in the works of creation. . . . Such facts loudly proclaim principles which science has not yet discussed, but which paleontological researches place before the eyes of the observer with increasing persistency; I mean the relation of the Creation to the Creator."

Nothing that occurred during the rest of Agassiz's life caused him to revise what he had said. He died in 1873. Harvard appointed three professors to replace him. Nine years later, in a scientific journal Agassiz had founded, his successor in the chair of geology published a paper describing the Ice Age as a myth. "The so-called glacial epoch . . . so popular a few years ago among glacial geologists may now be rejected without hesitation," the article concluded. "The glacial epoch was a local phenomenon."

.   .   .

West of Cleveland, the terrain became increasingly flat. High outcrops disappeared, but now and again a blocky strip of rock would run along the road like a retaining wall—a glimpse of what underlay the surrounding fields. Berea sandstone. Bedford shale. Columbus limestone. "You could map this state at sixty miles an hour," Anita said. For some distance, the soil over the rock was fine glacial till—ground rock flour and sand—and then among white farms we moved out upon a black-earth plain where drainage ditches did the work of streams: a world of absolute level, until recently the bottom of a great lake. The limestone had formed in the clear salt sea of middle Devonian tropical Ohio. Eventually, the sea disappeared. Two eons later, ice slid over the limestone and, retreating, left a body of fresh water that included all of what is now Lake Erie and was twice as large.

[197]

"We wouldn't be able to feed this country the way we do if much of it had not been glaciated," Anita said. "South of the glaciers, ancient weathering removed soluble minerals and left a rather inert soil behind. After a couple of decades of planting, you need tremendous fertilizer additions there. This glacial stuff is full of unweathered mineral material— fresh-ground rock. And under it is limestone, which is what they *put* on fields. When early settlers came through here and saw no trees, they moved on to places like Missouri, beyond the glacial limit, and they missed some great farmland. In Egypt, they used to get fresh minerals with every flood, but those morons built the Aswan High Dam and stopped the floods. They're starving themselves out and making a salt pan of the delta."

We were crossing the northern extension of the Cincinnati Arch and had reached the edge of the Michigan Basin, features of the subsurface structure, invisible and unexpressed in the black level surface of silts and clays. In tropical Ohio, the arch had at one time held back a large piece of the retreating sea. As the isolated water slowly concentrated and eventually disappeared, it left Morton's salt and U.S. gypsum. It left even more limestone. It left dolomite, anhydrite—components of what is known as the evaporite sequence. North off the interstate, we went through Gypsum, Ohio, on Sandusky Bay, and on to the lake port Marblehead, where we boarded the Kelleys Island ferry. "VISIT HISTORIC GLACIAL

GROOVES," said a sign beside the ticket booth, and soon, for a stiff toll, we were beating into an even stiffer wind, which was tearing the caps off the waves of Lake Erie. Kelleys Island is about four miles offshore, and other cars on the ferry were stuffed with a month's worth of groceries. A hundred and twenty people live there, year around, on four and a half square miles, and as we drove across the island we passed stone houses with red and black boulder walls—jaspers and amphibolites plucked up by the ice and brought south from the Canadian Shield.

Kelleys Island stands high because it is a piece of the Cincinnati-Findlay structural arch. While the Wisconsinan ice sheet was excavating the Great Lakes, reaming out whole networks of streams and carrying away the prominent features of their valleys, it bevelled but could not destroy the resistant structural arch. An engulfed ridge stood up from the bottom of the primal lake. With the weight of the ice gone, all of northern America slowly rebounded. A large part of the water gradually drained away, leaving Kelleys Island dry in the air, sixty feet above the level of Lake Erie.

We passed the island cemetery, its names recorded in limestone. We came to the north shore, where the beginnings of a quarrying operation had revealed how the ice had cut its tracks into the rock. "GLACIAL GROOVES STATE MEMORIAL." It was as if a giant had drawn his fingers through an acre of soft butter. The grooves were parallel. They were larger than the gutters of bowling lanes. Aggregately, they suggested the fluted shafts of Greek columns. Their compass orientation was northeast-southwest—the established glide path of the moving ice. Nowhere had we seen or would we see more emphatic evidence of continental glaciation, with the obvious exception of the Great Lakes themselves. "If you were to hydraulically flush northern Ohio—wash off the soil from the bedrock—you'd see a hell of a lot of these grooves," Anita said. "In several hundred years, these won't be here. Limestone is soft enough to be grooved and hard enough to resist weather for a few hundred years. In shale, grooves like these would go quickly. The ice, carrying boulders in its underside—carrying those amphibolites and red jaspers in the people's houses—tore the hell out of this island. When Agassiz saw things like this, he went bananas."

There have been glacial geologists, even in the late twentieth century, who have believed that such impressive grooves were gouged by boulders rolling in the Flood. Exceptions notwithstanding, Louis Agassiz's theory of continental glaciation, like the theory of plate tectonics, achieved with extraordinary swiftness its general acceptance in the world. As Thomas Kuhn has demonstrated in "The Structure of Scientific Revolutions," when a novel theory becomes relatively established it defines the patterns of amplifying research for many years and even centuries—until a new theory comes along to overturn the old, until an Einstein appears, outreaching the principles of Newton. Conceivably, the theory of plate tectonics will one day experience a general reformation. The theory of continental glaciation seems less prone to grand revision. The sun itself seems as likely to be banished from the center of the solar system as the ice from the Pleistocene continents. The ice made Lake Seneca, Lake Cayuga—all the so-called Finger Lakes, of western New York—cutting them into stream valleys in exactly the manner in which it cut the fjords of Patagonia, the fjords of Norway, Alaska, and Maine. After the ice quarried the huge quantities of Canadian rock that it dumped in the United States, it melted back and filled the quarries with new Canadian lakes—hundreds of thousands of Canadian lakes. A sixth of all the fresh water on earth is in Canadian ponds, Canadian streams, Canadian rivers, Canadian lakes. In Greenland, Antarctica, and elsewhere, a much greater quantity of fresh water—four times as much—is still imprisoned as ice, leaving precious little fresh water for the rest of the world.

Our *Époque Glaciaire* has by now been illuminated by more than a century of expanded research. Glacial outwash has been identified at the mouth of the Mississippi, six hundred miles from the terminal moraine—a suggestion of the power and the volume of the rivers that melted from the ice. Where the land tilted north and the meltwaters pooled against the glacial front—and where waters were trapped between moraines and retreating ice—gargantuan lakes formed, such as Glacial Lake Maumee, the one of which Lake Erie is all that remains. Lake Michigan is all that remains of Glacial Lake Chicago. Lake Ontario is all that remains of Glacial Lake Iroquois. Lake Winnipeg, Lake Manitoba, the Lake of the Woods are among the remains of a glacial lake whose bed and terraces, stream deltas and wave-cut shores reach seven hundred miles across Saskatchewan, Manitoba, and Ontario, and down into the United States as far as Milbank, South Dakota. With the exception of the Caspian Sea, this one was larger than any lake of the modern world. It was the supreme lake of the American Pleistocene—Glacial Lake Agassiz.

Cold air flowing off the ice sheets caused such heavy precipitation when it encountered warm and humid air to the south that whole regions there filled with water, too. The basins of Nevada became lakes and the ranges among them were islands. Lake Bonneville filled a third of Utah. Huge lakes grew in the Gobi Desert, in Australia's Great Artesian Basin, in various lowlands across North Africa. There were forests in the Sahara, as fossil pollen shows, and networks of flowing streams. Their dry channels remain.

In North America, where the ice started to go back about twenty thousand years ago, the first vegetation to spring up behind it was tundra. Carbon 14 can date the fossil tundra. The dates, particularly in the East, show a slow, and then accelerating, retreat.

After five thousand years, the front was still in Connecticut. In another twenty-five hundred years, it crossed the line to Canada. Human beings, living on the tundra near the ice, perforce were inventive and tough. Culture, in part, was a glacial effect. In response to the ice had come controlled fire, weapons, tools, and fur as clothing. Creativity is thought to have flourished in direct proportion to proximity to the glacier—an idea that must infuriate the equatorial mind. The ice drew back from Britain a geologic instant before the birth of Shakespeare. The fossils of *Homo sapiens* have never been found in sediments older than the Ice Age.

In the way that scenes of vanished mountains can be inferred from their debris, the vision of continents covered with ice came straightforwardly to Agassiz as the product of reasoning carried backward from evidence through time. It is one thing to say that the ice was there, quite another to say how it got there. If the origin of mountains is sublimely moot, so is the origin of the ice. Characteristically, the prime mover is not well understood. The ice did not come over the world like a can of paint poured out on the North Pole. It formed in places well below the Arctic Circle, and moved out in every direction—including north—until cut off by the unsupportive sea. Geologists call these places spreading centers, the same term they use for the rifted boundaries where plates tectonically divide. To the question "Why did the ice form?" they can answer only with speculation. The phenomenon is obviously rare. A pulsating series of ice sheets seems to have been set up in the discernible history of the world roughly once every three hundred million years. It happens so infrequently that it must be the result of coinciding circumstances that could not stand alone as explanations. There are components fast and slow. The atmosphere has been gradually cooling for sixty million years. Possibly this is explained by the great orogenies that have occurred during that time—the creation of the Rockies, the Andes, the Alps, the Himalaya—and the volcanism that is associated with mountain-building. Volcanic ash in the stratosphere reflects sunlight back into space. In any case, the essential requirement is a cool summer. A little snow from one winter must last into the next. Every forty thousand years, the earth's axis swings back and forth through three degrees. Summers are cooler when the earth is less tilted toward the sun. The sun, for that matter, is not consistent in the energy it produces. Moreover, the relative positions of the sun and the earth, in their lariat voyage through time, vary, too—enough for subtle influence on climate. Carbon dioxide also affects climate, and the amount of carbon dioxide in the atmosphere is not constant. Somewhere in such a list, which runs to many items, lie the simultaneous events that set the ice to growing. The change they bring is not at first dramatic. So critical is the earth's temperature that a drop of just a few degrees will cause ice to form and spread. A cool summer. Unmelted snow. An early fall in some penarctic valley.

An overlap of snow. A long winter. A new cool summer. An enlarged residue of snow. It compacts and
recrystallizes into granules, into ice. Because it is
white, it repels the sun's heat and helps cool the air
on its own. The process is self-enlarging, unstoppable,
and once the ice is really growing it moves. Clear
bands form near the base, along which the ice shears
and slides upon itself in horizontal layers like the
overthrust Appalachians. The thermal output of the
earth melts a thin film of water on the glacier bottom,
and the ice slides on that, too. Thrust sheets made of
rock also slide on water. The lower part of a glacier
is plastic, the upper part brittle—like the earth's
moving plates and the plastic mantle beneath them.
Where the brittle glacier surface bends, it cracks into
crevasses, into fracture zones, as does the brittle
ocean crust (the Clarion Fracture Zone, the Mendo-
cino Fracture Zone). Such fractures are everywhere
in the rock of continents, too. In fact, the ridged-and-
valleyed surface of almost any flowing glacier is remarkably similar to the sinuous topography of the
deformed Appalachian mountains. The continental
ice sheet moves toward the equator and keeps on
going until it cannot stand the heat. At the latitude
of New York City, generally speaking, the ice melts
as fast as it advances, and thus it goes no farther, and
leaves on Staten Island its terminal moraine. Ocean
temperatures will have dropped because of the cold,
and therefore the oceans are providing less snow to
feed the ice. On all fronts, the ice retreats—not necessarily to disappear. The climate warms. The oceans

warm. The snow pack thickens in the Great North Woods. A glacier spreads again. Once the pattern is set, the rhythm is relatively steady. For us, the ice is due again in ninety thousand years.

.   .   .

We ran on through Ohio on the bed of the great former lake, Kelleys Island far behind us. Where there had been sandspits reaching into the water, with sandy hooks at their tips, there were farm buildings standing on the dry spits—the high prime ground, a few feet higher than the surrounding fields. Now, at spring plowing time, these things were visible as they would not be for a year again.

And then we went off the lakebed and up into roadcuts of vetch-covered till among the kettles, kames, and drumlins, the Wabash Moraine, the New England landscape of glacial Indiana. "This would be a good place for a golf course," Anita remarked. "If you want a golf course, go to a glacier." We left the interstate for a time, the better to inspect the rough country. "I grew up in topography like this— in Brooklyn," Anita said. "I didn't know what bedrock meant. You could plot the limit of glaciation in New York City by the subway system. Where it's underground, it's behind the glaciation. Where it's in the moraine and the outwash plain, it's either elevated or in cuts in the ground."

Back on I-80—and running now on a pitted outwash plain, now on a moraine—we crossed the St. Joseph River. Anita's thoughts were still in Brooklyn. "My father died twenty years ago," she said. "When he was a little boy, his mother told him that if he ever ate food with his yarmulke off he would be struck dead. When she wasn't looking, he lifted his yarmulke and ate a spoonful of cereal. He didn't die. He quit believing. His faith was shaken."

There was a gold dome on our left—like an egg resting in a bed of new green canopy leaves. It was the supreme roof of the University of Notre Dame. "They're on outwash," Anita said in passing, and returned to her reminiscences. "Religious prejudice in any form is despicable," she went on. "In Brooklyn, when the Jehovah's Witnesses tried to sell me *The Watchtower* I'd say, 'I'm illiterate.' If they persisted, I'd say, 'Let me tell you about *my* God.'"

Again the St. Joseph River intersected the highway, and we ran on through grass-covered roadcuts of a kame complex, and soon through others in a recessional moraine, locally called the Valparaiso Moraine. A road sign suggested the proximity of Valparaiso. "Where do they get a name like that in a lacklustre place like Indiana?" Anita said, and we swung out an exit for the Indiana Dunes.

They are higher than the highest dunes of Cape Cod, and they are lined up in rows four deep along the shore of Lake Michigan—longitudinal dunes, transverse dunes, parabolic dunes. Glacial effects. On our map of Indiana, three of them were called mountains. They were covered with sand cherries, marram grass, cottonwoods, jack pines, junipers, and bluebells, except where the wind that made them had returned to them later to tear great blowouts in their sides. On foot, we approached the base of Mt. Tom. Staring upward, Anita said, "Look at the size of *that* son of a bitch." We climbed to the summit—to a view that might have pleased Balboa, had he been fond of power plants. There was a power plant to our left, a power plant to our right—Gary, Michigan City. Chicago was a shimmer of structures up the lake. Chicago was under water until two thousand years ago. The southern rim of Glacial Lake Chicago was the Valparaiso Moraine. As the lake level dropped, it left the makings of the Indiana Dunes. They are wind-built sands picked up from glacial till, so fresh that under a glass they are seen to be jagged. "It's been ground up so recently it's like the sand of Coney Island," said Anita, looking at it through her hand lens. She held the lens to her eye and a palmful of sand close to the lens. "I see angular grains of red chert," she said. "I see little fragments of igneous rock. I see amphibolite and red jasper—like the stuff that cut the grooves on Kelleys Island. I see red iron-oxide-coated quartz grains. You can see right through it to the quartz. I see little pieces of carbon. I see green chert. I see a bug crawling through the sand."

We sat in the lee of the top of Mt. Tom and watched whitecaps running on the lake. "As everybody knows, there are sand dunes in the Sahara," Anita said. "As everybody does not know, there are also grooves in the Sahara like the ones on Kelleys Island. They were cut in the bedrock in the Ordovician, when the Sahara was in a polar position and the equator was in Montreal."

I asked how the Sahara had accomplished such a journey.

"It is possible that the entire crust and some of the mantle can move around the interior earth—a sphere within a sphere," she said. "You've got to change the position of the continental landmasses with respect to latitude through time. That I can't deny."

"Do you believe that India smacked Asia?"

"I don't know. I know very little about the geology there, so how could I believe it? To many problems, plate tectonics is not the only solution. Often, it's a lazy man's out. It's a way of saying, 'I don't have to think any further.' It's a way of getting out of a problem. The geology has refuted plate-tectonic interpretations time and again in the Appalachians. Geology often refutes plate tectonics. So the plate-tectonics boys tend to ignore data. The horror is the ignoring of basic facts, not bothering to be constrained by data. It's like some modern art—done by people who throw paint at canvas and have never learned the fundamentals. Amateurs. Jumping into professionalism. Thinking they can get away with it. There are a lot of people out there in the profession

THE PALEOZOIC ERA

340 MILLION YEARS

MILLIONS OF YEARS BEFORE THE PRESENT	System / Period		Stage — EUROPE	Age — NORTH AMERICA
230				
	PERMIAN		TATARIAN	OCHOAN
			KAZANIAN	GUADALUPIAN
			KUNGURIAN	
			ARTINSKIAN	LEONARDIAN
			SAKMARIAN	WOLFCAMPIAN
280				
	PENNSYLVANIAN	CARBONIFEROUS	STEPHANIAN	VIRGILIAN
				MISSOURIAN
			WESTPHALIAN	DESMOINESIAN
				ATOKAN
310				MORROWAN
	MISSISSIPPIAN		NAMURIAN	CHESTERIAN
			VISEAN	MERAMECIAN
				OSAGEAN
345			TOURNAISIAN	KINDERHOOKIAN
	DEVONIAN		FAMENNIAN	CHAUTAUQUAN
			FRASNIAN	SENECAN
			GIVETIAN	
			COUVINIAN	ERIAN
			EMSIAN	ONANDAGAN
			SIEGENIAN	ORISKANYAN
			GEDINNIAN	HELDERBERGIAN
395				CAYUGAN
	SILURIAN		LUDLOVIAN	
			WENLOCKIAN	NIAGARAN
			LLANDOVERIAN	MEDINAN
435			ASHGILLIAN	CINCINNATIAN
			CARADOCIAN	TRENTONIAN
	ORDOVICIAN		LLANDEILIAN	BLACKRIVERAN
			LLANVIRNIAN	CHAZYAN
			ARENIGIAN	CANADIAN
			TREMADOCIAN	
500			DOLGELLIAN	CROIXIAN
			FESTINIOGIAN	
			MAENTWROGIAN	
	CAMBRIAN		MENEVIAN	ALBERTAN
			SOLVAN	
			CAERFAIAN	WAUCOBAN
570				